Principles and Standards for School Mathematics Navigations Series

NAVIGATING

through REASONING

and PROOF

in

GRADES 9–12

Maurice J. Burke
Jennifer Luebeck
Tami S. Martin
Sharon M. McCrone
Anthony V. Piccolino
Kate J. Riley

Maurice J. Burke
Grades 9–12 Editor

Peggy A. House
Navigations Series Editor

NATIONAL COUNCIL OF
TEACHERS OF MATHEMATICS

Copyright © 2008 by
The National Council of Teachers of Mathematics, Inc.
1906 Association Drive, Reston, VA 20191-1502
(703) 620-9840; (800) 235-7566; www.nctm.org

Navigating through reasoning and proof in grades 9-12 / Maurice J. Burke
... [et al.].
    p. cm. -- (Principles and standards for school mathematics
navigations series)
 Includes bibliographical references.
 ISBN 978-0-87353-604-2
 1. Proof theory--Study and teaching (Secondary)--United States. 2.
Logic, Symbolic and mathematical--Study and teaching (Secondary)--United
States. I. Burke, Maurice Joseph.
QA9.54.N38 2008
510.71'2--dc22
                        2007047049

The National Council of Teachers of Mathematics is a public voice of mathematics education, providing vision, leadership, and professional development to support teachers in ensuring mathematics learning of the highest quality for all students.

Printed in the United States of America

TABLE OF CONTENTS

CONTENTS OF THE CD-ROM

Introduction

Table of Standards and Expectations, Process Standards, Pre-K–Grade 12

Applet

Pythagorean Hexagon Construction

Blackline Masters and Templates

All blackline titles listed above, plus the following templates:

Geodot Paper

Centimeter Grid Paper

Supplemental Activities

The News

Read All about It! (blackline master)

Making Sense of the News (blackline master)

Solutions

Flooding a Water World

Keeping It Legal (blackline master)

Other Realms, Other Regions (blackline master)

Solutions

Readings from Publications of the National Council of Teachers of Mathematics

From Exploration to Generalization: An Introduction to Necessary and Sufficient Conditions

Martin V. Bonsangue and Gerald E. Gannon

Mathematics Teacher

Is It Always True? From Detecting Patterns to Forming Conjectures to Constructing Proofs

Elizabeth George Bremigan

Mathematics Teacher

Using Conjectures to Teach Students the Role of Proof

Rhonda L. Cox

Mathematics Teacher

About This Book

Navigating through Reasoning and Proof in Grades 9–12 discusses ideas and provides activities for building on and extending students' earlier work in problem solving and reasoning to include proof. In grades 9–12, students move from reasoning simply to obtain solutions to reasoning to justify ideas and establish the validity of results. They discover that justifying conjectures is essential in all five content areas of school mathematics—number and operations, algebra, geometry, measurement, and data analysis and probability.

Each of the five chapters in *Navigating through Reasoning and Proof in Grades 9–12* offers activities that draw on one of these five content strands. For each activity, the text specifies goals and materials and presents a discussion and suggestions for assessment. Accompanying activity sheets, along with solutions to the problems posed, are located in the appendix of the book. An icon in the text (see the key in the margin) signals all the reproducible blackline pages. These pages, as well as the solutions, also appear on the CD-ROM that accompanies the book. Thus, you can make copies of the activity sheets for your students' use from the book or by printing them directly from the CD-ROM.

A second icon in the text identifies all the supplemental materials that appear on the CD-ROM, including a number of pieces from publications of the National Council of Teachers of Mathematics for your professional development. The ideas and activities in the book give only glimpses of the possibilities for meeting the expectations of the Reasoning and Proof Standard. The journals of the National Council of Teachers of Mathematics suggest many other appropriate activities and approaches.

To enrich students' understanding, the CD also presents an applet, Pythagorean Hexagon Construction, and two additional activities, The News and Flooding a Water World (the latter also appeared in *Navigating through Number and Operations in Grades 9–12* [Burke et al. 2006]).

Throughout, margin notes supply teaching tips as well as pertinent statements from *Principles and Standards for School Mathematics* (NCTM 2000). A third icon flags these quotations, which highlight the fundamental notion that students should master mathematical processes and see mathematics—including proof—as an integrated whole.

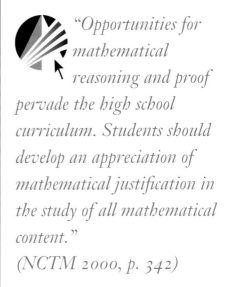

"Opportunities for mathematical reasoning and proof pervade the high school curriculum. Students should develop an appreciation of mathematical justification in the study of all mathematical content."
(NCTM 2000, p. 342)

Key to Icons

Principles and Standards

CD-ROM

Blackline Master

Three different icons appear in the book, as shown in the key. One signals the blackline masters and indicates their locations in the appendix, another points readers to supplementary materials on the CD-ROM that accompanies the book, and a third alerts readers to material quoted from *Principles and Standards for School Mathematics.*

NAVIGATIONS SERIES

GRADES 9–12

NAVIGATING *through* REASONING *and* PROOF

Introduction

Principles and Standards for School Mathematics (NCTM 2000, p. 342) calls for students in grades 9–12 to engage in four phases of reasoning and proof. All students are to—

- recognize reasoning and proof as fundamental aspects of mathematics;
- make and investigate mathematical conjectures;
- develop and evaluate mathematical arguments and proofs;
- select and use various types of reasoning and methods of proof.

These phases compose a cycle of exploration, conjecture, and justification (see figure 0.1). Almost all the activities in this book take students through this cycle, which reflects the fundamental habit of mind of the mathematician and points to the broader meaning of "reasoning and proof" in the context of *Principles and Standards*.

The phases of the cycle—exploration, conjecture, and proof—are interconnected and can occur in any order. Exploration includes investigating or questioning patterns or relationships. Conjecturing encompasses formulating, modifying, and extending hypotheses or generalizations. Justification involves validating, refuting, or otherwise explaining patterns, relationships, and conjectures.

Sometimes an exploration with numbers, figures, symbols, or events can lead to the development of one or more general statements, or conjectures, about their relationships or properties. When students are challenged to provide *irrefutable* evidence for these statements, they

Fig. **0.1.**
The reasoning and proof cycle (adapted from Martin, McCrone, and Pulley [2000])

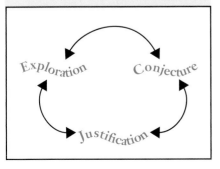

1

"Being able to reason is essential to understanding mathematics. By developing ideas, exploring phenomena, justifying results, and using mathematical conjectures in all content areas and – with different expectations of sophistication – at all grade levels, students should see and expect that mathematics makes sense."
(NCTM 2000, p. 56)

"Over the years of schooling, as teachers help students learn the norms for mathematical justification and proof, the repertoire of the types of reasoning available to students – algebraic and geometric reasoning, proportional reasoning, probabilistic reasoning, statistical reasoning, and so forth – should expand. Students need to encounter and build proficiency in all these forms with increasing sophistication as they move through the curriculum."
(NCTM 2000, p. 59)

encounter the need for justification as well as for valid methods of establishing the truth or falsity of claims. Once students are convinced of the truthfulness of a conjecture, they seek to justify their conclusions. Their search for a justification can lead to an explanation of the situation and can provide insight into why a relationship holds or does not hold.

Similarly, the process of justification can lead to new observations by heightening students' awareness of the assumptions in a problem and what might result from modifying those assumptions. Hence, a justification can lead to further exploration and new conjectures. It is through reasoning and reflection that one makes the leap from phase to phase in the cycle, in any order. Indeed, it is through such mathematical argumentation that students develop more sophisticated notions of what is meant by *irrefutable*.

Such is the nature of the cycle of reasoning and proof. It is a cycle that occurs in problem solving in each of the five content strands discussed in *Principles and Standards*—number and operation, algebra, geometry, measurement, and data analysis and probability.

Principles and Standards addresses the processes as well as the content of school mathematics. Like the other Process Standards, the Reasoning and Proof Standard is a vehicle for helping students learn mathematics and understand it more deeply. The focus of this Standard is on explanation, underscoring the importance of answering the question, "Why is this true?" Responding to this question is indispensable for making sense of mathematical phenomena.

High school mathematics curricula frequently relegate proof to a geometry course. Such curricula do not give students a balanced understanding of the "norms of mathematical justification and proof" (NCTM 2000, p. 59) that pertain in all the different content strands. *Principles and Standards for School Mathematics* explicitly intends for teachers to help students learn these norms as they apply to each strand. Understanding the use of reasoning and proof in each helps students recognize their essential roles in learning in all the strands.

One of the most challenging aspects of helping students learn the mathematical norms for reasoning and proof is nurturing the development of mathematical logic and mathematical language. The activities in the book do not focus on these two important achievements, but all of them do offer opportunities for the teacher and students to critique the logical form of the mathematical arguments that students develop in their investigations and to negotiate more precise mathematical meanings for the terminology and notations in use.

The development of logical reasoning has been the subject of a great deal of research. Numerous studies have focused on children's understanding of inference and proof from prekindergarten through grade 12. *Principles and Standards* encourages teachers to use activities like those in this book to nurture that logical development. However, for such activities to be effective, teachers must be aware of the level of sophistication of their students' logical reasoning. Posing questions throughout the activities can help reveal students' misconceptions.

To gain an idea of the complexity of the task of developing logical reasoning in the context of the activities in this book—or, indeed, any mathematical activity—you can engage your students in The News, an activity that appears on the accompanying CD-ROM. This extra activity illustrates the fact that mathematicians' norms for logical reasoning should not be taken for granted in the high school years. With modified tasks drawn from research studies on logical reasoning, the activity also illustrates how teachers can put research into practice to help uncover aspects of their students' reasoning that can affect their performance in contexts calling for reasoning and proof.

This book devotes a chapter to ideas and activities in each content area, thus securing an opportunity to illustrate the norms of justification and proof that are appropriate in all five content strands. This organization also allows for emphasizing the various types of reasoning encouraged by *Principles and Standards*. Working through activities from all the chapters can help students become aware of similarities and differences among the norms.

Chapter 1 illustrates the cycle of reasoning and proof in the learning of number and operation concepts. Activities demonstrate the suitability of using number and operation contexts to engage students in mathematical reasoning and proof. They highlight the role of counter-examples in mathematical reasoning. In addition, they expose the strengths and weaknesses of visual and symbolic proofs for "explaining" or making sense of phenomena in number and operations.

Chapter 2 illustrates the power of algebraic representation and reasoning in generalizing and proving results. By using the algebraic tools of variable and operation to model the mathematical phenomena introduced in the activities, students explain why certain multiplication algorithms work, discern the common structure of 3×3 magic squares, and categorize classes of functions according to their rates of change.

Chapter 3 addresses the issue of moving from exploration to proof in the context of geometric construction software, such as the Geometer's Sketchpad or Cabri Geometry, whose compelling representations can make certain conjectures seem irrefutable. The challenge in such a context is to find an explanation for the results that the technology so persuasively displays. Often, the technology does not reveal any clues as to why a conjecture is true. Activities in chapter 3 also emphasize the importance of reflecting on one's proofs as part of the reasoning and proof cycle.

Chapter 4 focuses primarily on probabilistic reasoning in the contexts of finite, countably infinite, and uncountably infinite sample spaces. Here, the reasoning and proof cycle leads students to simulations that yield empirical sampling distributions in which statistical reasoning reveals the significant trends and establishes plausible conclusions. Each activity directs students to theoretical approaches that substantiate the empirical results.

Chapter 5 presents activities in which the reasoning and proof cycle leads students to a deeper understanding of the relationship between the area and perimeter of rectangles, as well as to insights into the limit aspects of optimization problems and a better sense of the measurement process itself. These activities conclude with challenges to students to prove their conjectures.

Throughout, this book attempts to integrate many of the "Principles" of *Principles and Standards for School Mathematics*. The reasoning

The News, a supplemental activity on the CD-ROM, is based on research in syllogistic reasoning by L. J. Chapman and J. P. Chapman (1959).

"High school students should be able to present mathematical arguments in written forms that would be acceptable to professional mathematicians. The particular format of a mathematical justification or proof, be it narrative argument, "two-column proof," or a visual argument, is less important than a clear and correct communication of mathematical ideas appropriate to the students' grade level."
(NCTM 2000, p. 58)

"Reasoning and proof should be a consistent part of students' mathematical experience in prekindergarten through grade 12. Reasoning mathematically is a habit of mind, and like all habits, it must be developed through consistent use in many contexts." (NCTM 2000, p. 56)

and proof cycle, which weaves the activities together, lends itself to the task of achieving these Principles. It is our hope that the teachers who use this book will gain insights that they can incorporate into their own teaching and that they and their students will experience the satisfaction that reasoning and proof bring to all mathematical activities.

NAVIGATING *through* REASONING *and* PROOF

Chapter 1
Reasoning and Proof in Number and Operations

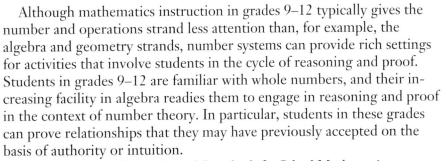

> "Students should understand that having many examples consistent with a conjecture may suggest that the conjecture is true but does not prove it, whereas one counterexample demonstrates that a conjecture is false."
> (NCTM 2000, p. 345)

Although mathematics instruction in grades 9–12 typically gives the number and operations strand less attention than, for example, the algebra and geometry strands, number systems can provide rich settings for activities that involve students in the cycle of reasoning and proof. Students in grades 9–12 are familiar with whole numbers, and their increasing facility in algebra readies them to engage in reasoning and proof in the context of number theory. In particular, students in these grades can prove relationships that they may have previously accepted on the basis of authority or intuition.

As elaborated in *Principles and Standards for School Mathematics* (NCTM 2000), the Number and Operations Standard expects all students to understand numbers, ways of representing numbers, relationships among numbers, and number systems. Students in grades 9–12 must develop particular skills and understanding related to number and operations. Among these are the abilities to—

- compare and contrast the properties of numbers and number systems …; and
- use number-theory arguments to justify relationships involving whole numbers. (NCTM 2000, p. 290)

These are the aspects of the Number and Operations Standard on which this chapter will focus in connection with reasoning and proof.

From kindergarten through grade 12, students have many experiences in which identifying a number pattern is not tied to an explanatory context that enables them to judge the validity of the pattern. Such experiences may lull students into believing that inductive reasoning is trustworthy and sufficient for generating convincing arguments. Surprising situations, such as the breakdown of the seductively simple numerical pattern in the activity Regions of a Circle, can serve to cast doubt on the reliability of inductive reasoning and persuade students of the need for general, deductive arguments when establishing truths in mathematics.

Chapter 1 presents two activities—Probing Products and Regions of a Circle. In Probing Products, students examine products of consecutive whole numbers as well as the squares of even and odd numbers. In Regions of a Circle, they investigate the maximum number of non-overlapping regions that can be formed by drawing all possible chords connecting n points identified on a circle.

Probing Products leads students to prove that the product of any two consecutive whole numbers is an even number. In the process, they generate and evaluate visual proofs. Be prepared for some students to think of these "proofs without words" as less than valid. At the other extreme, even more students may accept their validity without question. Such a spectrum of responses can allow you to discuss and negotiate with your students the requirements of a convincing mathematical argument. You should be prepared to encourage both skepticism about a proposed argument and appreciation for an argument that is valid.

Principles and Standards emphasizes that students should see mathematics as eminently reasonable. However, sometimes what appears to be reasonable turns out to be wrong. In such situations, the value of proof is very evident. Regions of a Circle provides such an experience. The students discover patterns and make conjectures as they explore a particular sequence of numbers associated with cutting a circle into regions. When challenged to justify a compelling conjecture that arises in the activity, students discover counterexamples and are forced to reconsider the limitations of inductive reasoning from a few examples.

Probing Products

Goals

- Generalize and justify patterns in products of consecutive whole numbers and in squares of even and odd numbers
- Connect products to figurate numbers
- Create and evaluate visual proofs

Materials and Equipment

For each student—

- A copy of the activity sheet "Probing Products"
- Two or three sheets of geodot paper (template available on the CD-ROM)
- (Optional) Access to a calculator for exploring conjectures

Discussion

In Probing Products, students explore relationships between factors and products represented numerically, algebraically, and in arrays of dots. They look for patterns in the products of consecutive whole numbers and discover that the products are always even. Students investigate the connection between the squares of whole numbers and *triangular numbers* (see the margin and fig. 1.1), and they prove that the squares of even numbers are always divisible by 4.

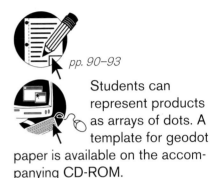

pp. 90–93

Students can represent products as arrays of dots. A template for geodot paper is available on the accompanying CD-ROM.

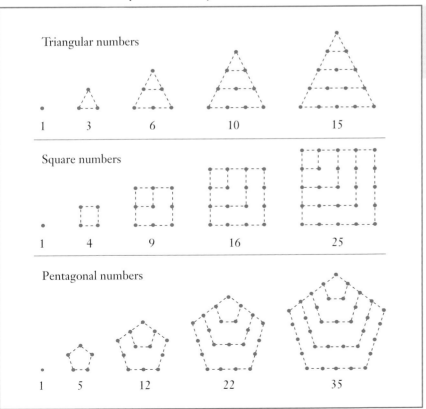

Triangular numbers

1   3   6   10   15

Square numbers

1   4   9   16   25

Pentagonal numbers

1   5   12   22   35

Fig. **1.1.**

The first five triangular, square, and pentagonal numbers

Triangular numbers are figurate numbers. A figurate number can be represented as a regular geometrical arrangement of dots (see fig. 1.1). Figurate numbers offer rich opportunities for exploring and justifying numerical relationships. For additional information on figurate numbers, see the following Web sites:

- http://mathworld.wolfram.com/ FigurateNumber.html
- http://math.youngzones.org/ FigurateNumbers.html

Students can use arrays of dots to represent the relationship between a product and its factors, and other relationships, geometrically. They can simplify their explorations of number patterns and relationships in the activity by working with geodot (square dot) paper, which shows rows and columns of dots. Figure 1.2, for example, illustrates how students can use dot arrays to demonstrate that the sum of the first n odd numbers is n^2. Give each student two or three sheets of geodot paper along with a copy of the activity sheet.

Fig. 1.2.

Working with dot arrays to show that the sum of the first n odd numbers is n^2

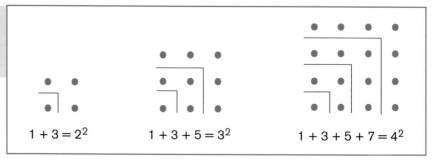

$$1 + 3 = 2^2 \qquad 1 + 3 + 5 = 3^2 \qquad 1 + 3 + 5 + 7 = 4^2$$

In the activity, the students use arrays of dots to create visual arguments demonstrating that—

- the product of consecutive whole numbers is even (step 2);
- the square of an even number is divisible by 4 (step 3);
- when 1 is subtracted from the square of an odd number, the difference is divisible by 8 (step 4).

They use triangular numbers in step 5 to make a visual representation of the last fact.

The visual arguments that students create in this activity provide experience with informal proof and an opportunity to discuss differences between formal and informal proof. Students often miss the subtle point that although some elements of a diagram are specific to the depicted situation, other elements are not and can serve as the basis for a general argument. For example, the construction in figure 1.3a, which shows a 6×6 array of dots divided into a 3×3 array of sets, with four dots in each set, demonstrates that any square of an even number is divisible by 4. Although the diagram is necessarily specific, an inference may be made to the general situation of a square with sides of length $2k$.

Fig. 1.3.

Diagrams that use a 6×6 dot array to demonstrate that if n is even, n^2 is divisible by 4

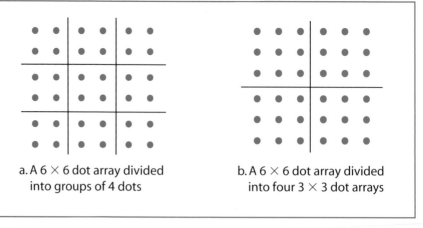

a. A 6×6 dot array divided into groups of 4 dots

b. A 6×6 dot array divided into four 3×3 dot arrays

The fact that the vertical and horizontal dimensions of the array are even numbers can act as the basis for an alternative visual justification of this result (see fig. 1.2b). This diagram guarantees, for example, that a $2k \times 2k$ array of dots can be divided in half both vertically and horizontally, thereby creating four $k \times k$ arrays, each with k^2 dots. This visual proof parallels a symbolic proof of the result $(2k)^2 = 4k^2$.

Assessment

Visual proofs provide valuable opportunities for you to discuss with your students the nature of proof and what counts as a convincing argument. Such proofs also afford occasions to address the relationship between informal representations and formal proof. Experiences with visual proofs can help clarify this important distinction for the students, at the same time enabling them to participate in setting standards for reasoning and proof in the classroom. Their work with and responses to visual proofs can also give you a better understanding of their conceptions of what constitutes a valid and general proof of mathematical propositions.

In step 1 of the activity, the students obtain the products of ten pairs of consecutive whole numbers, look for patterns, make a conjecture in "if … then" form about the products, and construct an argument to demonstrate the truth of the conjecture. Monitor the conjectures that your students make about the patterns in the products of consecutive whole numbers. Some students will immediately conjecture that the product is always even and then successfully complete an algebraic justification.

However, you may have a student who conjectures, for instance, that the product $n \times (n + 1)$ is always twice the triangular number T_n. This student might be quite challenged to write a convincing argument to demonstrate that the conjecture will always be true. Because the triangular numbers (see fig. 1.4) are so pervasive in mathematical applications that students encounter, it would be worthwhile, after the activity is completed, to ask the entire class to create dot paper representations to prove that the nth triangular number, clearly seen to be $1 + 2 + 3 + \ldots + n$, is equal to $\dfrac{n(n+1)}{2}$. Figure 1.4 provides a visual basis for this equality.

Not only will this task give you an opportunity to assess your students' grasp of the essential ideas of the activity, but it will also allow them to establish a very important result about numbers.

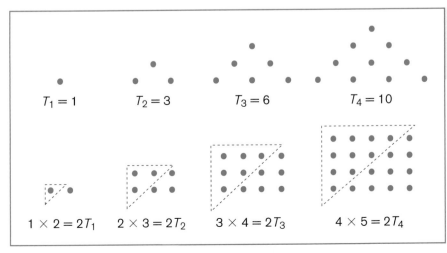

Fig. **1.4.**

A representation using dot arrays to show $n(n + 1) = 2T_n$

$T_1 = 1$ $T_2 = 3$ $T_3 = 6$ $T_4 = 10$

$1 \times 2 = 2T_1$ $2 \times 3 = 2T_2$ $3 \times 4 = 2T_3$ $4 \times 5 = 2T_4$

Steps 3(*a*) and 4(*a*), respectively, ask for conjectures about the square of an even number and the square of an odd number. The students may come up with various conjectures that are related to the ideas that n^2 is divisible by 4 if n is even or that $n^2 - 1$ is divisible by 8 if n is odd, though the conjectures may not state these facts directly. Steps 3(*b*) and 4(*b*) do propose these ideas explicitly. However, before your students proceed to these steps, consider leading a class discussion to connect their general conjectures from steps 3(*a*) and 4(*a*) to the specific propositions in steps 3(*b*) and 4(*b*).

In 3(*b*), students must prove that the square of any even number is divisible by 4, and in step 4(*b*) they must prove that the square of any odd number is one more than a multiple of 8. Figure 1.5 gives a visual proof of the latter result. You should encourage both visual and algebraic approaches to this proposition and assess whether or not students make use of the proposition that they proved in step 1—that the product of consecutive numbers, $n(n + 1)$, is even. For example, if $2k + 1$ is any odd number, then $(2k + 1)^2 = 4k^2 + 4k + 1 = 4[k(k + 1)] + 1$. However, by the proposition proven in step 1, $k(k + 1)$ is always even and hence equals $2h$ for some whole number h. Thus, $(2k + 1)^2 = 4[2h] + 1 = 8h + 1$. If a whole number n is odd, n^2 is one more than a multiple of 8.

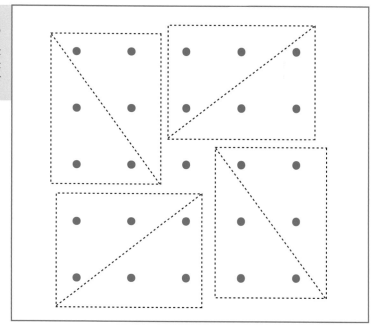

Fig. **1.5**

A visual demonstration of the proposition that one less than the square of an odd number is eight times a triangular number

Once the students have completed the activity and have demonstrated that the product $n(n + 1)$ is equal to $2T_n$, they will be in a position to create a different visual demonstration of the proposition that the square of an odd number $2k + 1$ is one more than eight times the kth triangular number. They can rewrite $2k + 1$ as $k + (k + 1)$ and then represent the square of the odd number $2k + 1$ as $[k + (k + 1)]^2$, as shown in fig. 1.6. They can divide the square, whose side length is $k + (k + 1)$, into four congruent rectangles, as shown, each with dimensions k by $(k + 1)$. The students can then apply the discovery that you helped them make earlier, that the product $n(n + 1)$ is equal to $2T_n$. This fact allows them to conclude that the square of the odd number $k + (k + 1)$ is equal to $4(2T_k) + 1$, or $8T_k + 1$.

Navigating through Reasoning and Proof in Grades 9–12

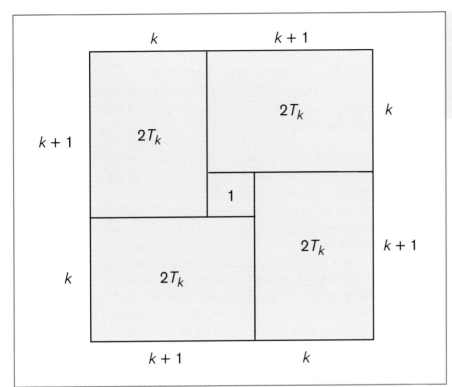

Fig. **1.6.**

A visual representation of the proposition that the square of the odd number $2k + 1$ is equal to $8T_k + 1$, given that the product $k(k + 1)$ is equal to $2T_k$

It is very important for students to understand that once a proposition has been proven, it can be used without proof in the validation of other propositions. This idea deserves special emphasis in the activity.

A variety of nice extensions can help you assess your students' understanding of the essential ideas in Probing Products. Each can give students the opportunity to explore, make conjectures, and justify or refute their conjectures. You can make these extensions more open-ended than the situations explored in the activity.

For example, you could ask students to find patterns in the products of consecutive odd or even numbers. Or you could have them explore the results when a square number is removed from the next larger square number, as in figure 1.7. Direct them to generalize their discoveries to the case in which any square number is removed from any larger square number.

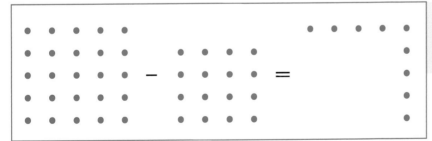

Fig. **1.7.**

A 4×4 square number subtracted from a 5×5 square number

You might also ask your students to explore the relationship between the square numbers and the triangular numbers. Figure 1.8 illustrates one such relationship. By separating a square dot array into two triangular arrays, one of which includes all the dots along the diagonal of the square array, one sees that every square number is the sum of two consecutive triangular numbers. In other words, the nth square number is $T_n + T_{n-1}$.

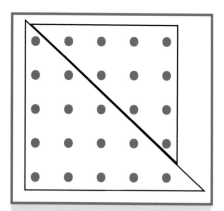

Fig. **1.8.**
The nth square number is $T_n + T_{n-1}$,
the sum of two consecutive
triangular numbers.

Extensions of Probing Products need not be limited to explorations of plane figurate numbers, such as triangular numbers or square numbers. Indeed, you might pose questions about the numerical relationships represented by arrangements of dots or sugar cubes in three-dimensional space. For example, what patterns do the students observe when a cube number is removed from the next larger cube number?

Principles and Standards recommends that students use number theory arguments to deepen their understanding of number systems and provide a basis for exploring other number systems. The algebraic representations required for some of the proofs in Probing Products demonstrate the need to work with general terms to create arguments that hold in all cases. They also allow students to make connections among algebra, number, and operation. Furthermore, the variety of formats that students encounter for proofs in this activity (symbolic, visual, verbal) gives them experience in using different modes of formal and informal reasoning.

When students engage in the process of exploration-conjecture-proof, they often make false conjectures. It is important for students to understand that refutation is an integral part of mathematical reasoning. When formulating generalizations and analyzing conjectures, they must appreciate the value of searching for counterexamples. The next activity, Regions of a Circle, gives students a chance to do just that.

Regions of a Circle

Goals

- Observe and generalize a number pattern
- Connect number patterns to geometric settings
- Establish the role and implications of counterexamples
- Establish the need for proof

Materials and Equipment

For each student—

- A copy of the activity sheet "Regions of a Circle"

For each group of three students—

- (Optional) Access to a computer with geometry construction software, such as the Geometer's Sketchpad, or a graphing calculator with Cabri Geometry

pp. 94–97

Discussion

In this activity, students count the maximum number of non-overlapping regions formed in a circle by drawing all possible chords that connect *n* points on the circle, for values of *n* ranging from 1 to 6. As the students examine these cases in sequence, they are likely to notice what appears be an emerging pattern.

The activity sheet provides diagrams for the cases of $n = 1$, $n = 2$, and $n = 3$ (see fig. 1.9). The students generate diagrams for $n = 4$, $n = 5$, and eventually $n = 6$, either by hand or by using technology (see the margin on the next page). The use of a high-tech tool such as dynamic geometry software can enable the students to obtain accurate data on which to base their conjectures as they explore the cases in which *n* is greater than 3. Such a tool can be very beneficial for this reason.

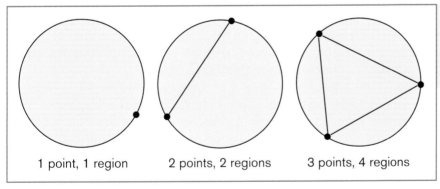

1 point, 1 region   2 points, 2 regions   3 points, 4 regions

Fig. **1.9.**

Non-overlapping regions in a circle, formed by drawing all possible chords to connect *n* points on a circle

After the students have considered the cases up to $n = 5$, the activity calls on them to formulate conjectures about the maximum number of interior regions that are created by chords connecting *n* points. They are asked to express their conjectures in formulas that let them compute the number of regions from the number of points on the circle, and they must explain why they think their formulas are true.

Students are quite likely to make the following conjecture: If *n* points on a circle are connected by all possible chords, then the maximum

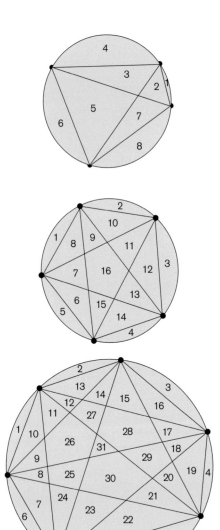

Three or more lines or line segments are *concurrent* if they intersect at a common point, which is called the *point of concurrence*.

n	1	2	3	4	5
A_n	3	6			

number of regions that may be formed in the interior of the circle is 2^{n-1}. They are likely to base their justifications for this conjecture on inductive reasoning, offering a statement such as, "I checked several examples and found a pattern."

Next, the students investigate the case of $n = 6$ to see if the pattern continues for six points. A hint on the activity sheet suggests that they take care to position the points on the circle in such a manner that the resulting chords maximize the number of interior regions. By making sketches, the students should eventually discover that to maximize the number of regions, they cannot allow three chords to be *concurrent* in the interior of the circle. Once the number of points on the circle exceeds five, if they let three chords intersect at a common point, they will miss an opportunity to create a region enclosed by those three chords.

The case of six points on the circle provides an opportunity to orchestrate a classroom discussion about the sufficiency of one counterexample to disprove a claim as well as the need for a general proof to substantiate a claim such as the one above. This case also offers a chance to examine the difference between an error and a counterexample. When all possible chords connect six points on a circle, one can easily err in counting the regions. Be sure that your students realize that they should all obtain the same number of regions if they have followed the rules of the problem. Otherwise, the significance of the counterexample might be lost, especially on students who think they have counted 32 regions instead of the maximum of 31.

It is important for students to understand that observing patterns and assuming that these will continue are useful steps in generating conjectures. Furthermore, though many conjectures turn out to be valid, they do not constitute sound mathematical arguments. To guarantee validity, an argument must provide a reasoned explanation of why a pattern will continue to hold in cases that are not specifically examined.

It turns out that the maximum number of non-overlapping regions generated in a circle by drawing all possible chords connecting n points on the circle is $\frac{1}{24}\left(n^4 - 6n^3 + 23n^2 - 18n + 24\right)$. Students with graphing calculators can easily discover this formula by using a quartic regression on the data corresponding to $1 \leq n \leq 5$. However, this method of generating the formula does not prove that the formula holds for the cases not specifically considered in the data set.

Steps 7–10 of the activity sheet lead students to the discovery that there are infinitely many polynomials exactly fitting the data in this problem for $1 \leq n \leq 5$. In these steps, the students consider a case in which values are known only for the first two terms, A_1 and A_2, in a sequence (see the table on the lower left in the margin). The students must complete the table and offer a formula for A_n. Step 7(c) reports that a student named Dania suggests that knowing values only for A_1 and A_2 "doesn't tell anything about the third term or any other terms in the sequence." Dania submits that the third term, "for all we know," could be the number of students in the class, or 25. She comes up with the formula $A_n = 8(n-1)(n-2) + 3n$, which gives correct values for A_1 and A_2 as well as 25 for A_3.

The students' work in steps 7–10 should help them discover that when they know only the values of A_1 and A_2, they cannot find a unique formula for the sequence $A_1, A_2, A_3, ..., A_n$. Infinitely many formulas exist that give the first two terms along with infinitely many possibilities for A_3. Help the students observe that the formula

$$A_n = k(n - 1)(n - 2) + 3n,$$

which is a general form of the formula that Dania suggests in step 8, yields $A_1 = 3$ and $A_2 = 6$, no matter what the value of k is. Furthermore, A_3 equals $9 + 2k$, and this expression has infinitely many possible values depending on one's choice of k.

Thus, if all someone knows about a sequence are the values of A_1 and A_2, he or she can say nothing for certain about the value of A_3. Students should discuss their answers to these questions. See if it makes sense to them to assert that if they know nothing about the process that generated the numbers A_1 and A_2, then why couldn't any number they choose be the value for A_3?

As their consideration of Dania's ideas will have suggested, validating the correct formula for the maximum number of regions is not an easy task. An activity on the CD-ROM provides a foundation for this proof, however. This activity, Flooding a Water World, reprinted from *Navigating through Number and Operations in Grades 9–12* (Burke et al. 2006), can serve as an extension of the activity Regions of a Circle. It develops counting strategies for determining the number of vertices, edges, and faces in a planar graph and proves Euler's formula for such graphs:

$$Faces = Edges - Vertices + 2$$

The students then use this formula to prove the formula that solves the problem in Regions of a Circle. You may also want to make your students aware of combinatorial arguments that do not use Euler's formula.

Assessment

Once your students have established counterexamples in the case of six (and perhaps seven) points, direct them to reflect on the justifications that they originally gave for their conjectures. It is important for students to articulate why these "justifications" turned out to be inadequate. Some will say that they just didn't consider enough cases. That may be true insofar as they were simply searching for counterexamples. But they must also realize that no amount of searching through cases would have proven the truth of the claim even if they had found no counterexamples.

In working through the mathematics curriculum for grades 6–12, students often find and extend a pattern in a finite sequence of numbers such as the apparent pattern that they find in the sequence in this activity. Nearly all the cases present a simple mathematical pattern that the students can find, as in this case (see the margin), where students are likely to identify the pattern $R_n = 2^{n-1}$. Furthermore, in nearly all cases, students are not asked to prove that the mathematical pattern actually works for all other values of n.

Not proving a conjecture for all n is sometimes appropriate in modeling applications, in cases where assuming that a carefully observed pattern continues leads to potentially useful consequences. However, in situations that do not involve modeling, omitting the proof might lead

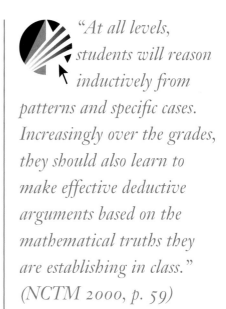

"At all levels, students will reason inductively from patterns and specific cases. Increasingly over the grades, they should also learn to make effective deductive arguments based on the mathematical truths they are establishing in class." (NCTM 2000, p. 59)

At the Web site http://www.cut-the-knot.org/Generalization/cuttingcircle.shtml, students can find another solution that also uses Euler's formula to find the maximum number of regions produced in a circle by drawing all possible chords that connect n points on the circle.

n	1	2	3	4	5
R_n	1	2	4	8	16

to the misconception that general formulas for sequences are justified by the fact that they fit all the data points specifically considered.

It is important that your students understand the need for a more general argument based on the process that produced the sequence of numbers in question—especially when presuming that there is only one appropriate extension of the pattern. In the case of Regions of a Circle, the students will find the process on which to base a general argument in the geometry of planar regions created by connecting points on a circle.

No matter how many numbers one might know in a finite sequence, one cannot say that those values by themselves determine any of the remaining values—not even the next value in the sequence. To assess your students' understanding of this important idea, you might ask them to evaluate what, if anything, they would need to know to find the next number in the following sequence of numbers:

n	1	2	3	4	...
G_n	2	4	6	?	...

Students should realize that they need hints that apply to all the terms of the sequence and the process or rules for generating it before they could even be sure of the value of G_4.

Conclusion

This chapter has focused on activities that highlight the process of reasoning and proof in the context of number and operations. The next chapter shifts the context to algebra while continuing to emphasize the reasoning and proof cycle.

NAVIGATIONS SERIES

GRADES 9–12

NAVIGATING *through* REASONING *and* PROOF

Chapter 2
Reasoning and Proof in Algebra

Algebra is a mathematical language of variables and operations, used to model real-world and mathematical situations. Algebraic models and representations include operations on variables that represent numbers, functions, sets, propositions, and other mathematical objects, as well as real-world phenomena and processes. Algebra supports reasoning and proof in many branches of mathematics. The skills and tools of algebra strengthen students' abilities to "visualize, describe, and analyze situations in mathematical terms" and "use a wide range of explicitly and recursively defined functions to model the world around them," as *Principles and Standards for School Mathematics* (NCTM 2000, p. 288) recommends.

The scenarios in this chapter's activities call algebraic reasoning into play. As the Algebra Standard for grades 9–12 suggests, the activities offer students experience in—

• identifying and codifying arithmetic patterns;

• modeling the properties of function graphs and other graphs;

• representing relationships between variables; and

• modeling change or other real-world processes.

By creating and working with an algebraic model—often a function and its attributes—students achieve a greater understanding of the situation or phenomenon that gave rise to the model. Mastery of algebra gives students knowledge, skills, and techniques that they can use to explore, conjecture, analyze, and prove propositions about the systemic properties of operations on variables. The study of algebra provides a rich context in which students can develop their skills in symbolic reasoning. Indeed, instruction that emphasizes reasoning can transform algebra from a procedural and formula-based study, focused on the ability to memorize, apply, and combine processes, to a study that calls for creative and original thinking.

By selecting appropriate algebraic contexts, teachers can encourage the practice and development of specific reasoning skills. Their students can—

- create, test, refute or prove *conjectures*—sometimes by examining all possible cases—as they explore and determine the general properties of algorithms and structures of numbers or operations;
- employ methods of *justification* and move from case-by-case analyses to more general arguments as they make and defend predictions about the changes in the behavior of algebraic expressions and functions over the domain of a variable;
- use *deductive reasoning* with algebraic processes that preserve the equivalence of expressions, such as factoring, expanding, or combining like terms in an expression, applying operations to both sides of an equation, and reexpressing functions.

The activities in this chapter offer several contexts that call on students to apply algebraic reasoning directly. The first activity, Algorithmic Aptitude, invites students to analyze patterns of errors in flawed computational procedures. They explore alternative algorithms, and invent and validate their own algorithms. The activity includes an algorithm that the students can validate by examining all possible cases. They also have a chance to analyze a given algorithm by constructing an algebraic argument of a more general type on the basis of place value (for example, any two-digit number ab is equal to $10a + b$).

In the second activity, Demystifying Magic Squares, the students reason and make representations to investigate the algebraic structure underlying the simple, finite system of numbers in the traditional 3×3 magic square puzzle. This time, in contrast to the strategies that they used to validate most of the cases in Algorithmic Aptitude, examining all possible cases is an efficient proof strategy for some of the propositions. However, the students will discover that using algebraic representations and processes allows them to make the best proofs for many propositions in the activity.

Linear, quadratic, and exponential functions provide the setting for pattern-based reasoning in the third activity, Grappling with Growth Rates. Here analyzing changes in linear, quadratic, and exponential functions leads the students to proofs about attributes of entire families of these functions. The graphing calculator provides an ideal platform for exploring and formulating conjectures about the growth rates of functions. However, the students must realize that checking every possible case is not possible. Proving that a pattern holds for an infinite number of possible cases calls for a more general approach. The methods of proof suggested in this activity are algebraic or geometric but rely on generalization in either case.

"Understanding change is fundamental to understanding functions."
(NCTM 2000, p. 40)

Algorithmic Aptitude

Goals

- Investigate nontraditional computational algorithms
- Diagnose and analyze arithmetic error patterns
- Express and verify algorithms in the general case

Materials and Equipment

For each student—

- A copy of the activity sheet "Algorithmic Aptitude"
- A graphing calculator

Discussion

By the time that students reach high school they should be developing mental shortcuts for performing basic calculations. The ability to perform mental arithmetic efficiently is both natural and necessary for a variety of tasks in daily life, such as determining a tip, calculating a discount, and comparing two billing plans. Problem solvers frequently invoke the distributive property to break into simpler steps such everyday mental calculations as estimating a 15 percent tip or multiplying a number by 21. Algorithmic Aptitude gives students a chance to gain a better understanding of the nature of algorithms by analyzing error patterns in nontraditional algorithms for whole-number computation.

In preparation for this activity, make sure that your students understand the basic meaning of *algorithm*. An algorithm is a step-by-step procedure for completing a task. Discuss the existence of algorithmic processes in everyday life—for example, ordered routines for putting on shoes and socks or for mixing ingredients and baking a peach pie. Emphasize the importance of properly sequencing the steps and providing clear instructions. Pose questions to make the point—for instance, "If you put on shoes before socks, do you get the same result as when you put on socks before shoes?

This activity emphasizes the importance of verifying algorithms by using algebraic reasoning. The students encounter an algorithm that "works" for some specific cases but not in general. They discover the need for caution and proof. They realize that the validity of the algorithms that they learned in earlier grades rests not on an arbitrary claim—"It's the law!"—but instead on properties of number, operation, and place value.

Analysis of error patterns in arithmetic serves as a means of demonstrating this idea. By examining correct and incorrect results produced with the same algorithm, students can see that knowing that an algorithm works for several examples is not enough to verify the algorithm. One must justify the structure of the algorithm. The second and fourth examples in figure 2.1 are correct, but the first and third examples reveal a subtle error pattern—the student apparently assumes that any subtraction involving a zero results in zero.

pp. 98–101

Many unique algorithmic shortcuts can be found in Short-Cut Math *(Kelly 1984).*

An algorithm *is a step-by-step procedure for completing a task.*

$$
\begin{array}{cccc}
147 & 624 & 527 & 805 \\
-20 & -323 & -304 & -201 \\
\hline
120 & 301 & 203 & 604
\end{array}
$$

Fig. **2.1.**

An error pattern in subtraction: any subtraction involving a zero results in a zero

Even when alternative algorithms are correct, they are not necessarily simpler than traditional procedures. Students should recognize that methods have different advantages—some methods are more efficient than others, some are better models for the operation, and some are very creative but not very practical. The activity presents a "teener multiplication" algorithm for multiplying two "teener" numbers—that is, whole numbers between 11 and 19. Although the application of this algorithm is limited, it gives problem solvers a very efficient way to compute the product of "teener" numbers mentally.

For example, when using the teener algorithm to multiply 13 by 17, a problem solver simply—

- adds 3 (the ones digit from the 13) to 17;
- shifts the resulting sum, 20, one decimal place to the left, getting 200;
- adds on the product of 3 times 7 (the ones digits from 13 and 17); and
- correctly obtains 221 as the product of 13 times 17.

With a calculator to speed up the process, the students could validate this algorithm by exhaustively examining the 81 possible ordered pairings of the numbers in the set. However, hints in the activity should enable them instead to validate the algorithm algebraically.

Representing two teener numbers as $10 + b$ and $10 + n$, where b and n are whole numbers from 1 to 9, the teener product is equivalent to $(10 + b + n)(10) + (b \times n)$. By expanding this expression and comparing the result to $(10 + b)(10 + n)$, the students verify that the products from the nontraditional algorithm and the standard algebraic algorithm are equivalent:

$$(10 + b + n)(10) + (b \times n) =$$
$$100 + 10b + 10n + bn =$$
$$(10 + b)(10 + n).$$

Assessment

The last step in the activity calls on the students to create a non-traditional algorithm for a simple computational procedure. Ask your students to write a set of instructions for their algorithm, provide several examples, and use algebra to justify that their procedure will work in all cases. Encourage them to include a description of how their method reduces intermediate steps, minimizes calculation, or otherwise improves on traditional methods.

Alternatively, you might ask your students to create their own flawed computational methods for an arithmetic operation. Or they could come up with a flawed algebraic procedure for a more complex situation—for example, an incorrect application of the distributive property or binomial multiplication. Challenge them to create a flawed method that produces correct answers in some cases. Ask them to demonstrate their flawed procedure with several examples that yield both correct and incorrect results, and have them write a clear and concise explanation of the steps involved in the procedure. Encourage your students to use mathematical language to describe the operations involved. Let them exchange papers and try to identify the error patterns in the procedures invented by their peers.

The activity Algorithmic Aptitude emphasizes reasoning in a variety of ways. Students look for patterns to identify computational errors, and they develop patterns to justify procedures for performing calculations. The activity's tasks encourage students to reason about the algebraic structures that underlie algorithmic procedures and discover the importance of going beyond an exhaustive examination of cases to justify mathematical procedures in a generalized form. The next activity, Demystifying Magic Squares, links algebraic reasoning, number sense, and problem solving in the context of exploring the algebraic structure of a familiar puzzle—a 3×3 magic square.

Demystifying Magic Squares

Goals

- Use algebraic representations and properties of numbers to identify patterns and verify properties of magic squares
- Apply reasoning by processes of elimination, examining cases, and other strategies to justify the structure of magic squares
- Communicate discoveries and justifications in algebraic terms
- Use contrapositive reasoning to rule out possibilities

Materials and Equipment

For each student—
- A copy of each of the following activity sheets:
  - "Demystifying Magic Squares"
  - "3 × 3 Magic Square Grid"
  - "3 × 3 Magic Square Numbers"
- A pair of scissors (to cut out the numbers on the activity sheet "3 × 3 Magic Square Numbers")
- (Optional) A calculator or access to a computer with spreadsheet software (for quickly constructing and calculating row, column, and diagonal sums of 3 × 3 number squares)

Discussion

For many years, mathematics teachers have presented magic squares to their students primarily as number puzzles or challenge problems. Rarely do teachers or students take the time to delve into the algebraic structure of the squares. The activity Demystifying Magic Squares provides students with one such opportunity. Like the preceding activity, Algorithmic Aptitude, this activity encourages students to link a system of arithmetic computations with its underlying algebraic structure and to verify the generality of that structure by using algebraic representations and arguments.

The activity defines a *magic square* as any square array of whole numbers with no two numbers alike and whose columns, rows, and diagonals all yield the same "magic" sum. For example, figure 2.2 shows a 3 × 3 magic square and a 4 × 4 magic square. These two magic squares fit the traditional definition of an $n \times n$ magic square, which restricts the allowable numbers to the set $\{1, 2, 3, \ldots, n^2\}$, with no repeats.

The activity begins by asking the students to verify the sums in a given 4 × 4 magic square. You can take advantage of this exercise to make sure that your students understand what a magic square is and appreciate the complexity of its structure. A 4 × 4 magic square has many more permutations than a 3 × 3 magic square does. The permutations are classifiable but too complex for exploration in one activity. You may want to challenge students to explore the structure of 4 × 4 magic squares as an extension of their work in the activity.

After introducing the students to a 4 × 4 magic square in step 1, the activity turns to an investigation of 3 × 3 magic squares. Given nine

pp. 102–6, 107, 108

In a *magic square*, each small square contains a whole number that is different from all the others. Adding the numbers in any row, column, or diagonal of a magic square always gives the same sum. An $n \times n$ magic square traditionally, though not always, uses each of the numbers 1 to n^2 exactly once, as in the examples in figure 2.2.

Fig. 2.2.

A 3 × 3 and a 4 × 4 magic square

8	1	6
3	5	7
4	9	2

1	15	14	4
12	6	7	9
8	10	11	5
13	3	2	16

consecutive whole numbers, the students can completely determine the possibilities for their arrangement in a 3 × 3 magic square with relative ease, making such a magic square an excellent tool for exploration and verification.

Step 2 challenges the students to complete a given 3 × 3 magic square (see the 3 × 3 magic square in fig. 2.2; the square in step 2 shows the numbers 1 and 2 in place, but the students must position the remaining numbers). Without knowledge of the "infrastructure" of a magic square, they are likely to use a guess-and-check method for placing the numbers. Working with the square grid and the number markers provided with the activity can facilitate the students' exploration of the possibilities.

Ask your students to describe how they proceeded, and note whether they followed a systematic process in their search for a magic square. One important outcome of this initial exploration in step 2 is that students should start to understand why the magic sum of each row, column, and diagonal must be 15. Adding all three rows of numbers together is equivalent to summing all the numbers in the 3 × 3 magic square. For example, for the magic square in step 2, 1 + 2 + 3 + 4 + 5 + 6 + 7 + 8 + 9 = 45. But the sum of all three rows is also three times the magic sum. Thus, since 45 is three times the magic sum, the magic sum is 15. In step 4, the students will prove this formally by using algebra.

Meanwhile, the questions in step 2 direct the students' attention to the possible placement of numbers in the center square. In a 3 × 3 magic square, the center number is unique in that it is linked with every one of the other eight numbers in a row, column, or diagonal sum. The students may also observe that they can never join 6, 7, or 8 with 9 in a row, column, or diagonal of a 3 × 3 magic square formed from the numbers 1 through 9. Otherwise, the sum for that row, column, or diagonal would be larger than 15. Because 9 can never appear with 6, 7, or 8 in any row, column, or diagonal, none of these four numbers can occupy the center square. Likewise, the students can never join 2, 3, or 4 with 1 in a row, column, or diagonal of such a magic square. Otherwise, the sum for that row, column, or diagonal would necessarily be less than 15. Hence, the number in the center square can never be 1, 2, 3, or 4. Therefore, by a process of elimination, the students will discover that the center square must contain 5.

In steps 3 and 4(*a*), the students explore the patterns and parameters of a successful 3 × 3 magic square. Some students will notice that in each case in which they succeed in building a magic square, the center square is a third of the magic sum and is the median of the set of nine numbers in the square. Students might also notice that the center square is always the average, or mean, of the nine numbers. For example, 5 is the mean of the numbers 1 through 9. Although the students can establish these conjectures for each of the specific sets of numbers that they succeed in arranging in a magic square, they need algebra to prove the conjecture in general. Step 4(*b*) gives the students the nine numbers in variable form, as in the magic square in the margin, and guides them to proofs of these conjectures.

In step 5, the students continue to consider the general 3 × 3 magic square, looking now for a relationship between the average of the nine

a	*b*	*c*
d	*e*	*f*
g	*h*	*i*

numbers in the square and the number in the center. Let T be equal to the total of all nine numbers and S be the square's magic sum. The students know that

$$T = (a + b + c) + (d + e + f) + (g + h + i),$$

the total of the sums of the three rows of the square. But the sum of each row is S, so $T = 3S$. Furthermore, the students know that dividing T by 9 gives the average of the nine numbers in the magic square. Dividing $T = 3S$ by 9 gives

$$\frac{T}{9} = \frac{3S}{9} = \frac{S}{3}.$$

Thus, the equation $T = 3S$ indicates not only that the magic sum is one third of the total of the nine numbers but also that the average of the nine numbers is one third of the magic sum.

To prove that the center number in a 3×3 magic square is the average of the nine numbers making up the square, and hence one-third of the magic sum, students must dig deeper into the algebraic structure of the square. They must identify all the information that they can about the center number e and its relationship with the magic sum S. They can represent this information conveniently in algebraic form:

$$S = a + e + i, S = b + e + h, S = c + e + g, \text{ and } S = d + e + f.$$

The challenge for students is to figure out how to combine this information to draw valid conclusions about e. Step 5(*b*) gives them a hint about how to do this: "Combine the equations $S = d + e + f$, $S = a + e + i$, $S = b + e + h$, and $S = c + e + g$ to show that e is one-third of the magic sum S." Students who observed that the magic sum was always three times the center number in the squares that they considered might not need this hint. Using addition to combine the equations involving S and e, the students can deduce that

$$4S = (a + e + i) + (b + e + h) + (c + e + g) + (d + e + f).$$

Rearranging terms will give them

$$4S = (a + b + c + d + e + f + g + h + i) + 3e,$$

or $4S = T + 3e$. However, since $T = 3S$, the students can conclude that $4S = 3S + 3e$, or $S = 3e$, or $\frac{S}{3} = e$. Thus, the center number e is the average of the nine numbers and one-third of the magic sum S.

In step 6, the students apply what they have learned to determine whether two given sets of nine numbers can be arranged in magic squares, and they create a set of nine numbers that cannot be arranged in a magic square. These exercises can offer opportunities for assessment (see below). They also lead smoothly into the activity's concluding steps (7–10), which guide the students through an examination of the structure of all 3×3 magic squares.

Armed with the knowledge that the number in the center square is one-third the magic sum, the students can verify that all 3×3 magic squares have the form shown in figure 2.3. Hence, any 3×3 magic square is generated by three whole numbers a, x, and y.

Students must be careful, however, in the way that they understand the generalization. If a 3×3 array of numbers is a magic square, then it fits the pattern shown in figure 2.3. However, every 3×3 number

"Students should reason in a wide range of mathematical and applied settings…. Algebra is conducive to symbolic reasoning. Students who can use many types of reasoning and forms of argument will have resources for more effective reasoning in every-day situations."

(NCTM 2000, p. 345)

Fig. **2.3.**

The general form of all 3×3 magic squares

$a + x$	$a - x - y$	$a + y$
$a - x + y$	a	$a + x - y$
$a - y$	$a + x + y$	$a - x$

square that fits the pattern is not necessarily a magic square. A 3×3 number square in which every number is equal fits the pattern but is not a magic square. With most mathematical generalizations, it is important to distinguish a statement from its converse. Be sure that your students are aware that verifying a statement does not imply that its converse is also valid.

Assessment

As suggested above, step 6 can help you assess whether your students have grasped the essential properties of 3×3 magic squares. This step can also demonstrate their ability to reason from the *contrapositive* of the statements that they have just proven.

Be sure that your students understand what a contrapositive is. Ask them to suppose that a statement "If P, then Q" is true for two propositions P and Q. Explain that the statement "If not Q, then not P," which is the contrapositive of the original statement, is also true.

How can students reason contrapositively to complete step 6? Consider the following example. The students now know that if nine whole numbers form a magic square, then the average of the nine numbers must be one of the nine numbers. The contrapositive must also be true: if the average of nine whole numbers is not itself one of the nine numbers, then the nine numbers must not form a magic square. The average of the nine numbers in step 6(*a*) (the set {1, 3, 5, 7, 8, 10, 12, 14, 16}) is $\frac{76}{9}$, which is not one of the nine numbers. Therefore, the nine numbers cannot form a magic square.

Some students will try to form the numbers into a magic square and conclude that it can't be done because nothing that they tried worked. Other students will first try to calculate the magic sum by applying the principle that this sum must be one-third of the total of the nine numbers. When they discover that this magic sum would have to be $25\frac{1}{3}$, they should conclude that adding three of the given whole numbers could never equal this magic sum.

To extend the activity or assess your students' understanding, you might ask each student to select a page from a conventional SMTWTFS calendar for one month of the year and box off an arbitrary 3×3 square of numbers. Then have them determine whether the nine numbers in their box can be rearranged to form a magic square. They should now know enough about the structure of 3×3 magic squares to succeed in rearranging the numbers in a magic square.

Once they have succeeded in making a magic square and have compared their results with their peers, you can give them an additional challenge. Ask them to prove that they can construct a magic square from every unique 3×3 square of numbers that they can box off in all possible arrangements of calendar numbers on a SMTWTFS calendar. The calendar offers a total of fifteen possibilities (see fig. 2.4).

Students could prove this claim by listing all fifteen possible 3×3 squares of numbers from a calendar and showing that each can be rearranged to form a magic square. Encourage your students instead to try to prove their results by using algebraic representations, as in figure

Be sure that your students understand that verifying a statement does not imply that its converse is also valid.

If "P implies Q" is a true statement for two propositions P and Q, then "Not Q implies not P" is also a true statement. "Not Q implies not P" is called the *contrapositive* of the original statement and has the same truth value.

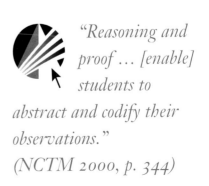

"Reasoning and proof … [enable] students to abstract and codify their observations."
(*NCTM 2000, p. 344*)

1	2	3
8	9	10
15	16	17

2	3	4
9	10	11
16	17	18

3	4	5
10	11	12
17	18	19

...

13	14	15
20	21	22
27	28	29

14	15	16
21	22	23
28	29	30

15	16	17
22	23	24
29	30	31

Fig. **2.4.**
Fifteen 3 × 3 number squares can be formed from all possible arrangements of the numbers 1–31 in a conventional SMTWTFS calendar.

2.5. The students should discover that if n is the smallest number in the square from the calendar, the magic sum is always $3n + 24$, and the number in the center of the magic square is always $n + 8$ (which also appears in the center of the number square taken from the calendar).

The activity Demystifying Magic Squares offers students a context in which they can prove the properties of a finite algebraic structure by using basic algebraic means—or at times by considering all possible cases. The next activity, Grappling with Growth Rates, encourages algebraic and occasionally geometric reasoning by asking students to verify and prove general relationships about sequences formed from linear, quadratic, and exponential functions.

Fig. **2.5.**
Using algebra to represent the rearrangement of numbers in any 3 × 3 number square from the calendar to form a magic square

12	13	14
19	20	21
26	27	28

Arranges to →

13	28	19
26	20	14
21	12	27

n	$n + 1$	$n + 2$
$n + 7$	$n + 8$	$n + 9$
$n + 14$	$n + 15$	$n + 16$

Arranges to →

$n + 1$	$n + 16$	$n + 7$
$n + 14$	$n + 8$	$n + 2$
$n + 9$	n	$n + 15$

Navigating through Reasoning and Proof in Grades 9–12

Grappling with Growth Rates

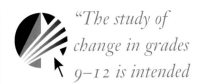

"The study of change in grades 9–12 is intended to give students a deeper understanding of the ways in which change in quantities can be represented mathematically and of the concept of rate of change."
(NCTM 2000, p. 305)

Goals

- Analyze changes in polynomial and exponential functions by using consecutive differences
- Use algebra to validate the observable patterns in tables of function values
- Use contrapositive reasoning to rule out possibilities

Materials and Equipment

For each student—

- A copy of the activity sheet "Grappling with Growth Rates"
- A graphing calculator

pp. 109–13

Discussion

The study of rates of change in functions is typically reserved for calculus courses. *Principles and Standards for School Mathematics* encourages teachers to introduce the study of rates of change much earlier. "If ideas of change receive a more explicit focus from the early grades on, perhaps students will … enter calculus with a stronger basis for understanding the ideas" (NCTM 2000, p. 40). The activity Grappling with Growth Rates provides an approach to the notion of change that uses precalculus methods and algebra.

This activity defines the growth rate function $\Delta f(x)$ corresponding to a function $f(x)$ as the difference in $f(x)$ for consecutive values of x:

$$\Delta f(x) = f(x + 1) - f(x)$$

The growth rate function $\Delta f(x)$ represents the amount of growth—a quantity that might be negative—in the values of the function $f(x)$ between x and $x + 1$. This quantity can be thought of as the "rate of change" of $\Delta f(x)$ between x and $x + 1$. In other words, the growth rate is the ratio $\dfrac{Change\ in\ f(x)}{Change\ in\ x}$, since by definition, "*Change in x*" in this case is 1.

For example, if $f(x) = x^2$, then $\Delta f(x) = (x + 1)^2 - x^2$, which simplifies to $\Delta f(x) = 2x + 1$. As shown in the table of values displayed in figure 2.6, $\Delta f(x)$ represents the consecutive differences of the sequence generated by $f(x)$ when x varies over the natural numbers. (The table displays only values corresponding to $1 \le x \le 6$.) For example, $\Delta f(5)$ is equal to 11:

$$\Delta f(5) = f(6) - f(5) = 36 - 25 = 11$$

In a graph of $f(x)$, the growth rate function $\Delta f(x)$ gives the values of the slopes of secant lines at points for $f(x)$. See figure 2.7. The slope of the secant line between $x = 1$ and $x = 2$ on the graph of $f(x) = x^2$ is

The notation "$\Delta f(x)$" for the growth rate function is read as "delta *f* of *x*."

Fig. **2.6**.

A partial table of values, showing
corresponding values of $f(x)$ and $\Delta f(x)$ for
$f(x) = x^2$, where x is a natural number

x	1	2	3	4	5	6
$f(x) = x^2$	1	4	9	16	25	36
$\Delta f(x)$	3	5	7	9	11	

$\dfrac{f(2) - f(1)}{2 - 1}$, which equals $f(2) - f(1)$, or $\Delta f(x)$. For comparison, figure 2.7 also shows the graph of the line tangent to $f(x) = x^2$ at $x = 1$. In calculus, the instantaneous rate of change of a function $f(x)$ at a point $x = n$ is defined to be $f'(n)$, the derivative function of $f(n)$, provided that this derivative exists. In the graph of $f(x)$, $f'(n)$ is the slope of the line tangent to the graph of $f(x)$ at $x = n$. As the comparison of the slopes of the secant line and the tangent line in figure 2.7 suggests, the growth rate function $\Delta f(x)$ gives precalculus students a way of evaluating the rates of change in functions at various points on their graphs without having to define the derivative function.

Fig. **2.7**.

The slope of the secant segment from
$A = (1, 1)$ to $B = (2, 4)$ is $\Delta f(1)$.

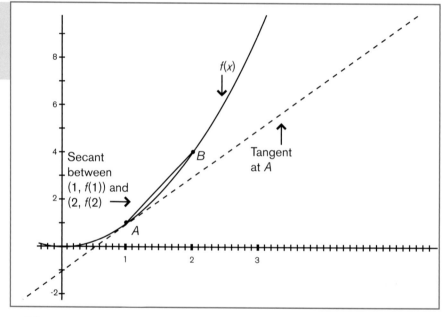

The activity encourages students to find growth rate functions by first examining tables of values of the functions over the domain of natural numbers. For example, by examining the consecutive differences shown in figure 2.6 for the function $f(x) = x^2$, the students can readily discover that the pattern in the sequence of values for $\Delta f(x)$ seems to be $2x + 1$. However, the idea emphasized in the activity Regions of a Circle in chapter 1 is worth stressing here: the fact that $2x + 1$ fits the first five terms in the sequence of values for $\Delta f(x)$ does not prove that it will continue to describe the later terms in the sequence. To prove that $2x + 1$ is the pattern for all terms in the Δf sequence, students must use algebra to investigate the general case of $\Delta f(x)$. That is, they must evaluate $f(x + 1) - f(x)$.

If students are familiar with the composition of functions, they can evaluate $f(x + 1) - f(x)$ directly. Otherwise, encourage them to create a table of values, insert the general cases $x = n$ and $x = n + 1$ into their table, and calculate $\Delta f(n)$. Figure 2.8 shows the table from figure 2.6 with the additions of $x = n$ and $x = n + 1$. By evaluating $(n + 1)^2 - n^2$, students prove that $\Delta f(x) = 2n + 1$, which matches the formula that they discovered empirically. It is interesting to note that figure 1.7 (see p. 11) suggests a visual demonstration of this particular algebraic proof on geodot paper.

The activity guides the students in generating several growth functions for linear, quadratic, and exponential functions and then asks them to interpret their results by creating and comparing graphs of $f(x)$ and $\Delta f(x)$. They should observe several patterns. First, when $f(x)$ is linear, $\Delta f(x)$ is a constant that is equal to the slope of $f(x)$. When $f(x)$ is quadratic, $\Delta f(x)$ is a linear function whose positive values generally, but not always, correspond to the points where $f(x)$ is increasing and whose negative values generally correspond to the points where $f(x)$ is decreasing. When $f(x)$ is an exponential function of the form $f(x) = a(b^x)$, $\Delta f(x)$ is also an exponential function. Furthermore, by considering the table of exponential values, the students should discover that $\dfrac{\Delta f(x)}{f(x)}$ appears to be a constant.

Step 6 in the activity gives the students an efficient process for creating the necessary graphs on their graphing calculators. This use of the graphing calculator illustrates the power of technology to help students explore and formulate algebraic conjectures. The graphing calculator allows students to enter a function $f(x)$ and then, without knowing the explicit formula for $\Delta f(x)$, quickly generate a graph of $\Delta f(x)$ and a table that juxtaposes the values of $\Delta f(x)$ with those of $f(x)$. In step 6, this feature of the calculator allows students to investigate numerous quadratic functions rapidly, readily observing that the corresponding growth rate functions always appear to be linear. Likewise, in steps 7 and 8, the students can quickly investigate numerous exponential functions and notice that the corresponding growth rate functions always appear to be exponential. However, the students must bear in mind the fact that such calculator investigations, though compelling, are not proofs.

The steps in the activity ask the students to justify their observations and generalizations. In the cases where $f(x)$ is an arbitrary linear function, students might argue that the slope of a line is always constant between two points on the line. Therefore $\Delta f(x)$, which measures the slope of the graph of $f(x)$ between points, must be constant. When $f(x)$ is a quadratic function, students might argue that $\Delta f(x)$ is a linear function because the pattern that they observed in all the cases that they tried was that $\Delta f(x)$ was linear. Encourage them to move beyond the empirically observed pattern, strengthening their argument by making a difference table for the general case $f(x) = ax^2 + bx + c$, just as they did

"Students should use technological tools to represent and study the behavior of polynomial, exponential, rational, and periodic functions, among others."
(NCTM 2000, p. 297)

Fig. 2.8.

Extending the table of values for $f(x)$ and $\Delta f(x)$ to include the general cases $x = n$ and $x = n + 1$

x	1	2	3	4	5	6	...	n	$n + 1$
$f(x) = x^2$	1	4	9	16	25	36	...	n^2	$(n + 1)^2$
$\Delta f(x)$	3	5	7	9	11		...	$(n + 1)^2 - n^2$	

for specific quadratic functions. Offer the same advice and encouragement when they attempt to validate their observations about the class of exponential functions. Figure 2.9 shows a difference table that includes the general case of the exponential function. To approach the general case successfully, the students need to be able to factor and simplify expressions involving exponents.

Step 9, the last step in the activity, gives students five consecutive terms in a function $p(x)$ (see the table of values in the margin). They must determine whether or not the function could be a linear function, a quadratic function, an exponential function, or none of these. This is another good setting for contrapositive reasoning. Students who have done the preceding activity will recall that contrapositive reasoning represents the following inference pattern where P and Q are propositions: Given that "If P, then Q" is a true statement, then the statement "If not Q, then not P" is also true. In step 9, the students know that if $p(x)$ is quadratic, then $\Delta p(x)$ is linear. If by graphing or otherwise inspecting some of the values of $\Delta p(x)$ students can show that $\Delta p(x)$ is not linear, then they can conclude that $p(x)$ is not a quadratic function. Likewise, $p(x)$ is not an exponential function, since the values of $\dfrac{\Delta f(x)}{f(x)}$ are not constant. Opportunities for contrapositive reasoning abound in mathematics. Be sure that your students understand that any time when they have proved a statement of the form "All A are B" and they discover that c is not a B, then they can conclude that c is not an A.

Assessment

One of the most difficult aspects of this activity is likely to be the use of function notation, especially where the composition of functions is required. For example, the definition of $\Delta f(x)$, given as $f(x + 1) - f(x)$, involves the composition of functions. Students often confuse this notation with multiplicative concepts. For instance, students sometimes think that $f(x + 1)$ is equivalent to $f(x) + f(1)$. The activity attempts to get students around such difficulties by having them include the variables n and $n + 1$ in their tables of values, as in figures 2.8 and 2.9. In any case, it is important to be watchful as your students represent their general arguments and proofs.

Many of the explorations in this activity are performed on calculators, thus producing no written record of what the students are actually observing. This circumstance presents an excellent opportunity for you

x	$p(x)$
1	2
2	16
3	54
4	128
5	250

The table of values in step 9

Fig. 2.9.

Making a table of values for the growth rate function $\Delta f(x)$ for the general exponential function $f(x) = a(b)^x$

x	1	2	3	4	...	n	$n + 1$
$f(x) = a(b)^x$	ab	ab^2	ab^3	ab^4	...	ab^n	ab^{n+1}
$\Delta f(x)$	$ab^2 - ab$ or $a(b-1)b^1$	$a(b-1)b^2$	$a(b-1)b^3$	$a(b-1)b^4$	...	$ab^{n+1} - ab^n$ or $a(b-1)b^n$	

Navigating through Reasoning and Proof in Grades 9–12

to promote writing in mathematics. Students should keep a record of what they do on a calculator, documenting results and sketching some of the graphs that they so quickly produce. Without a written record, you might have difficulty in assessing your students' grasp of specific concepts and connections. For instance, would you be sure that they understand that the growth rate function $\Delta f(x)$ provides an indicator of the slope of the graph of $f(x)$?

As discussed, your students' work in step 9 can help you assess whether or not they have grasped the main ideas in the activity. However, you can also assess their essential understanding by challenging them to demonstrate the general attributes of the growth rate functions corresponding to $f(x) = x^3$ or $f(x) = \dfrac{1}{x}$ —or indeed the functions $H(n)$ and $R(n)$ for the fees charged by the High Rentz and Rising Sums hotels in step 1.

Conclusion

The activity Grappling with Growth Rates allows students to conduct a small but meaningful exploration of a pervasive theme in mathematics—rates of change. Like the other activities in this chapter, this final activity engages students in the cycle of exploration, conjecture, and proof or refutation in an algebraic context. In addition, it illustrates how technology can facilitate the exploration of a conjecture but usually does not provide the generality of scope needed to prove a result. Many of these same themes arise in the activities in the next chapter, which focuses on reasoning and proof in geometry.

"As high school students study several classes of functions and become familiar with the properties of each, they should begin to see that classifying functions as linear, quadratic, or exponential makes sense because the functions in each of these classes share important attributes. Many of these attributes are global characteristics of the functions."
(NCTM 2000, p. 300)

NAVIGATING
through
REASONING
and PROOF

Chapter 3
Reasoning and Proof
in Geometry

"Along with making and investigating conjectures, students should learn to answer the question, Why does this work?"
(NCTM 2000, p. 58)

Geometry offers students countless opportunities to explore, discover, make conjectures, refine them, and prove or refute claims. Dynamic geometry software has been a boon to teachers and students alike, enriching and expanding these opportunities and motivating problem solvers to move beyond their conjectures to answer the larger question, "Why does this work?" *Principles and Standards for School Mathematics* (NCTM 2000) recommends that students in grades 9–12 use their reasoning abilities to confirm or refute conjectures through deductive arguments or by finding counterexamples.

Reasoning and proof are in fact central to the goals that *Principles and Standards* sets out for geometry in grades 9–12. This chapter focuses on several expectations of the Geometry Standard, including recommendations that instruction in high school offer all students opportunities to—

- explore relationships (including congruence and similarity) among classes of two-and three-dimensional geometric objects, make and test conjectures about them, and solve problems with them;

- establish the validity of geometric conjectures using deduction, prove theorems, and critique arguments made by others;

- investigate conjectures and solve problems involving two- and three-dimensional objects represented with Cartesian coordinates. (NCTM 2000, p. 308)

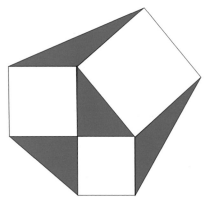

A Pythagorean hexagon

Chapter 3 presents three activities, each of which engages students in the reasoning and proof cycle. Assisted by technology, the students generate very compelling conjectures with comparative ease, but they may find the proofs of these conjectures to be more elusive. The activities help students make the transition from conjecture to proof.

In the chapter's first activity, Savvy Surveyor, students explore the possibilities for splitting a triangle into smaller triangles of equal areas by using line segments. The conjecture that any median divides a triangle into two triangles of equal area is not hard to prove. However, dividing the triangle into three triangles of equal area is not so easy, as the students discover as they move through the activity.

A successful proof does not always mark the end of the reasoning and proof cycle. In fact, a proof is often the beginning of a new exploration. Geometry in grades 9–12 is replete with such instances. In the chapter's second activity, A Hex on Pythagoras, students once again encounter an equal-area problem. The activity introduces the idea of a Pythagorean hexagon—a hexagon formed by constructing a right triangle, drawing squares on all three sides, and connecting vertices of the squares to form a hexagon (see the margin). The students conjecture that the four triangles in the hexagon are equal in area. Hints about how to manipulate the Pythagorean hexagons shed light on a strategy for proving this conjecture. However, once the students succeed in the proof, the activity calls on them to analyze their work to see if certain assumptions that they made in solving the problem were really necessary. This leads students to generalizations that in turn motivate them to undertake new explorations.

Questioning one's assumptions is also at the center of the chapter's third and final activity, Centroid of Gravity. This activity exhibits the power of algebraic representations in geometric reasoning and proof. What seems to be a modest discovery about line segments leads to an unexpected result about circles and arbitrary sets of points in the Cartesian plane. By examining a proof to determine the assumptions on which it rests and considering the effects of modifying those assumptions, the students engage in a critical aspect of mathematical reasoning and the "looking back" process that Polya (1945, pp. 14–15) describes. Centroid of Gravity once again reminds the students that proof leads to further exploration and discovery.

Savvy Surveyor

Goals

- Analyze properties—particularly the areas—of two-dimensional geometric shapes
- Make and test conjectures about two-dimensional geometric objects
- Develop valid mathematical arguments about geometric relationships involving the centroid of a triangle

Materials and Equipment

For each student—
- A copy of the activity sheet "Savvy Surveyor"
- Several sheets of grid paper (template available on the CD-ROM)
- A pair of scissors

For each group of three students—
- (Optional) Access to a computer with a dynamic geometry construction package (such as Cabri Geometry or the Geometer's Sketchpad)

pp. 114–16

You can print grid paper for your students' use in this activity from the template "Centimeter Grid Paper," which appears on the accompanying CD-ROM.

Discussion

The activity Savvy Surveyor guides students in exploring ideas and investigating geometric relationships involving the centroid of a triangle. It focuses students' attention on the medians of a triangle and challenges them to organize their knowledge about both medians and centroids by answering two larger, entwined questions: "Why does this work?" and "What else does this reveal?"

The activity opens with a scenario about an eccentric rancher whose land is a triangular tract. The old rancher has hired the student, supposedly known to be a savvy land surveyor, to divide the land into two smaller triangles of equal area for the rancher to give to his two children. In step 1, the students look for a segment that will divide the triangular region into two triangular parts of equal area. Infinitely many segments will partition a triangle into two equal parts, but the rancher has imposed a limiting condition: the two parts of the triangular whole must themselves be triangular. Hence, the partitioning segment must start at a vertex of the original triangle. The students should determine fairly quickly that a *median*—a segment connecting a vertex of a triangle to the midpoint of the triangle's opposite side—serves as an appropriate boundary. But why does the median divide a triangle into two smaller triangles of equal area, no matter what triangle or vertex one starts with? The students must come up with a proof and present their findings to their classmates.

Whether students discover the median by using dynamic geometry software or by drawing their own triangles on grid paper, a proof of the result follows directly from the calculation of the area of a triangle. For

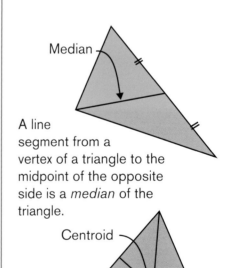

Median

A line segment from a vertex of a triangle to the midpoint of the opposite side is a *median* of the triangle.

Centroid

The *centroid* of a triangle is the point at which the triangle's three medians intersect. It divides the length of each median in a ratio of 2 to 1 from the vertex to the midpoint of the opposite side.

example, in figure 3.1, the median $\overline{AD}$ and the altitude $\overline{AE}$ have been drawn on $\triangle ABC$. Since D is the midpoint of $\overline{BC}$, and $\overline{AE}$ is the altitude for both $\triangle ABD$ and $\triangle ADC$, then the two triangles have the same base and height measurements, and hence, the same areas. Though the proof is basic, the result has many applications.

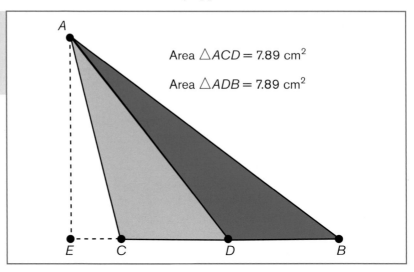

Area $\triangle ACD = 7.89$ cm^2

Area $\triangle ADB = 7.89$ cm^2

Fig. **3.1.**

The median $\overline{AD}$ divides $\triangle ABC$ into two parts of equal area.

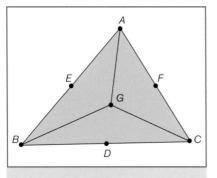

Fig. **3.2.**

$\triangle ABC$ with centroid G defining $\triangle ACG$, $\triangle ABG$, and $\triangle BCG$

In step 2, the students learn that the rancher may decide to keep a third of the land himself, so they now explore possibilities for dividing the rancher's triangular tract into three equal triangular regions. They consider a particular three-way division that the eccentric rancher likes (see fig. 3.2). The rancher is unsure whether this division produces equal regions, so the students must show that all three triangular regions in the map have the same area.

The students' work in this step reminds them that all three medians of a triangle meet at the *centroid*. This point of intersection is two-thirds of the way along each median from the vertex to the midpoint opposite the vertex. If the students have access to geometric construction software, they can use it to demonstrate to the rancher that the three medians of any triangle intersect at this common point. In any case, the students should investigate the conjecture that the three triangles formed by constructing segments from the centroid to the three vertices of the original triangle, as in the rancher's map, always have equal areas. Experiments with the software can make this conjecture seem extremely plausible to the students. Nevertheless, a proof of the fact that this method of dividing a triangle into three parts always produces triangles of equal area may elude them. The students are advised to look for clues in their construction of the centroid, but if they do not succeed in finding a proof at this stage, step 3 may show them a way.

Many times in mathematics, solving a simpler problem enables problem solvers to approach a more difficult problem successfully. For example, step 3 draws the students' attention to the fact that the construction of the centroid actually divides the original triangle into six smaller triangles (see fig. 3.3a). If the students can show that these six triangles (with areas a, b, c, d, e, and f) have the same areas, then they can readily prove the conjecture in step 2. Step 3 points them in this direction, hinting that they might start by proving that $a = b$, $c = d$, and $e = f$.

The key to successful completion of steps 2 and 3 is the basic result discovered and proved in step 1. In figure 3.3, the dotted segments from the centroid G are actually medians of the three triangles ABG, ACG, and BCG. Thus, they divide those triangles into two triangles of equal area, as discovered in step 1. Hence, $a = b$, $c = d$, and $e = f$ (see fig. 3.3b). Likewise, since median $\overline{AD}$ divides $\triangle ABC$ into two triangles of equal area,

$$a + b + c = f + e + d$$

(see fig. 3.3a), and with appropriate substitution,

$$a + a + c = e + e + c$$

(see fig. 3.3b), which implies $a = e$. By similar reasoning about median $\overline{CE}$, $a = c$. Putting the equalities together, we know that all six smaller triangles have the same areas.

Step 4 of Savvy Surveyor concludes the activity with an exercise that helps the students think back over their investigations and use their understanding of centroids and medians to solve a new problem. They are given two triangles, $\triangle ABC$ and $\triangle DEF$, with $\angle ABC$ supplementary to $\angle DEF$, $AB = DE$, and $BC = EF$ (see fig. 3.4). On the basis of their earlier work, they can say with certainty that the areas of $\triangle ABC$ and $\triangle DEF$ are equal, and they should be able to justify this conclusion with relative ease. This work should take them smoothly to a new generalization about triangles: If two sides of one triangle are congruent, respectively, with two sides of another triangle, and the included angles of these sides are supplementary, then the two triangles have the same area.

This generalization will prove useful in the next activity, A Hex on Pythagoras. The process of looking back and extending newly discovered ideas illustrates what Polya means when he says, "By looking back at the completed solution, by reconsidering and reexamining the result and the path that led to it, they [students] could consolidate their knowledge and develop their ability to solve problems.... [N]o problem whatever is completely exhausted" (Polya 1945, p. 14).

Assessment

For students of geometry, the discovery phase of the reasoning process is usually easier than the proof phase. Students must realize that the key to proving their conjectures often lies in the properties of the constructions that led them to make the conjectures in the first place. In

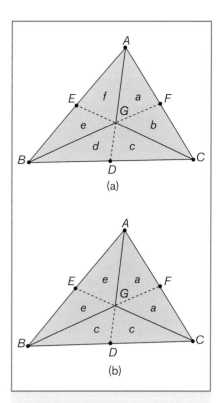

Fig. 3.3.

Six triangular regions created by the medians of $\triangle ABC$

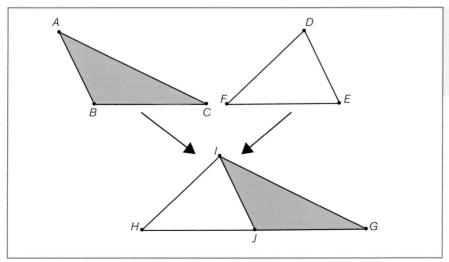

Fig. 3.4.
Two triangles, $\triangle ABC$ and $\triangle DEF$, with $\angle ABC$ supplementary to $\angle DEF$, $AB = DE$, and $BC = EF$, translated to form $\triangle GHI$

step 1, some students who discover that the median divides the triangle into two parts of equal area will not know where to begin the proof. Or they may try a proof startegy that is inappropriate, such as trying to prove that the two triangles produced by the median are congruent. You may need to suggest that students approach the task of proving the equality of the two areas by considering how they determine the area of a triangle. Then they can simply use the formula for the area: the area of a triangle is one-half the base times the height. When the students sketch the altitude and apply the formula directly, they should quickly see that both triangles have the same base and height.

Once your students have succeeded in completing step 1, you can check their understanding of the key ideas by having them draw two segments from a vertex so that the triangle is divided into three equal areas. This same principle can be used to divide the triangle into any number of triangles of equal area. When the students thoroughly understand the principle behind a theorem, they usually end up knowing how to prove several related theorems.

In step 2, most students will succeed in convincing themselves that the areas of $\triangle ABG$, $\triangle ACG$, and $\triangle BCG$ are equal (see fig. 3.2). However, most will struggle with the task of proving their conjecture. It is important to let your students search for reasons and share their thinking. You should monitor them closely and encourage them to exchange ideas before going on to step 3.

When the students move to step 3, you may need to help them observe that the dashed segments $\overline{GD}$, $\overline{GE}$, and $\overline{GF}$ (see fig. 3.3a) from the centroid are actually medians of the three triangles BCG, ABG, and ACG, respectively. Hence, the dashed segments divide each of the three triangles into two triangles of equal area (see fig. 3.3b). Again, this observation comes from a careful analysis of the construction and leads to the conjecture that the three triangles $\triangle ABG$, $\triangle ACG$, and $\triangle BCG$ have equal area.

The students begin work in step 4 by cutting out a copy of the triangle in step 1 (see the margin) to compare the two pieces they get when they cut the triangle along a median. Although they know that the areas of the two pieces are equal, it is important for them to dispel any notion that the two parts are congruent. Superimposing one of the triangles on the other does this very efficiently. It is also important that the students notice that the process of separating the two triangles can be reversed. That is, if two triangles can be joined at a common side so that the result is a single triangle with a median along the joined sides, then the two triangles have the same area (see fig. 3.4).

Have the students share the various ways in which they express this observation. Attempt to determine whether they are able to recognize statements that are essentially equivalent. For example, fig. 3.4 can be summarized by the following statement: "If $\angle ABC$ and $\angle DEF$ are supplementary, with $AB = DE$ and $BC = EF$, then the areas of $\triangle ABC$ and $\triangle DEF$ are equal." This proposition can be proven by showing that a triangle IGH can be constructed so that the triangles ABC and DEF are congruent, respectively, to triangles IJG and IJH formed by a median of $\triangle IGH$.

A goal of *Principles and Standards for School Mathematics* is to promote the teaching of mathematics in ways that allow students to make

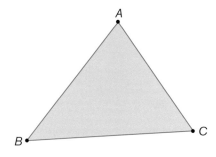

sense of what they are learning. To acquire a deep understanding of mathematical ideas, students need to have many experiences in seeking and finding explanations for patterns that they observe. *Principles and Standards* encourages classroom use of dynamic geometry for exploring a range of geometric examples and facilitating students' development of conjectures about geometric relationships. In the following activity, A Hex on Pythagoras, students use dynamic geometry software to explore and make conjectures about geometric relationships found in two-dimensional geometric shapes related to the Pythagorean theorem. Once again, close examination of the construction that leads to the conjecture provides clues for proving the result.

A Hex on Pythagoras

Goals

- Critique assumptions in mathematical arguments and generalize results
- Explore and analyze properties of two-dimensional geometric shapes
- Use dynamic geometry software to generate and test conjectures
- Develop valid mathematical arguments about geometric relationships

Materials and Equipment

For each student—

- A copy of the activity sheet "A Hex on Pythagoras"

For each group of three students—

- Access to a computer with a dynamic geometric construction package (such as Cabri Geometry or the Geometer's Sketchpad) or the applet Pythagorean Hexagon Construction (available on the CD-ROM)

pp. 117–18

Discussion

In A Hex on Pythagoras, students use dynamic geometry software to identify invariant relationships in a geometric figure that the activity refers to as a "Pythagorean hexagon." A Pythagorean hexagon is formed by first constructing what looks like a visualization of the Pythagorean theorem. After a square is drawn on each side of the triangle, line segments are added to connect vertices of the squares to create a convex hexagon.

In this activity, as in all others involving computer representations of invariant relationships, it is important for students to understand the distinction that the technology makes between a "drawing" and a "construction." For example, if a user produces a right triangle whose right angle he or she creates by dragging a vertex of a triangle until the angle measurement becomes 90°, then the user has made a "drawing." The triangle on the screen will cease to be a right triangle if the user continues to drag the vertex or one of the other vertices. If, however, the user "constructs" the right angle by selecting the perpendicular line option offered by the software, then the property of being a right triangle becomes an invariant attribute of the triangle on the screen, and this property is preserved when the user drags a vertex. Be sure that your students grasp the difference between a "drawing" and a "construction" before they begin work on the activity.

Step 1 asks the students to construct a Pythagorean hexagon, starting with a right triangle (see fig. 3.5). The applet Pythagorean Hexagon Construction, included on the CD-ROM that accompanies this book, provides a dynamic construction that students can use if they do not have access to commercial dynamic geometry software. However, if your students complete the construction themselves, make sure that they do not "draw" the right angles by using appearances or measurement tools to determine when an angle is 90°. The students should "construct" all right angles in the figure by using the perpendicular and parallel line options or the square-making tools provided with the software.

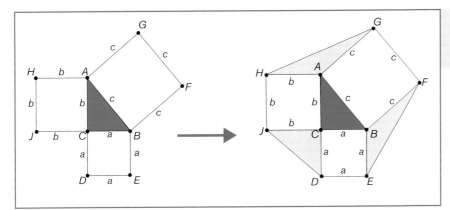

Fig. **3.5.**

Generating a Pythagorean hexagon

Step 2 directs the students to explore the features of a Pythagorean hexagon by using the software's measurement options and by dragging the vertices of △ABC to create different hexagons. The students are asked what relationships they notice. A hint suggests that the students measure the areas of the squares and triangles and the lengths of the segments and then look for relationships among these measurements.

Step 3 takes the students from an inspection of sample Pythagorean hexagons to a consideration of the properties of all such hexagons. The students now make and share conjectures about the invariant relationships that they discover. One conjecture that the students are likely to produce quickly is that the areas of the four triangles in the Pythagorean hexagon are always equal, no matter how the right triangle changes as its vertices are dragged. Indeed, this conjecture, which is the focus of the remainder of the activity, does not depend on △ABC being a right triangle. Any triangle will work, as the students will discover.

After the students have discussed their conjectures in step 3, step 4 asks them to prove that the areas of the four triangles are equal. A hint advises them to prove that each of the three external triangles in the hexagon has the same area as △ABC. You might suggest that each student in a group of three take a different external triangle and prove that it has the same area as △ABC.

If the students are familiar with the idea from trigonometry that figure 3.6a illustrates, along with the fact that the sine of an angle and the sine of its supplementary angle are equal (see fig. 3.6b), then the proof is straightforward. For example, in figure 3.5, the area of △EBF is $\frac{1}{2}ac(\sin\angle EBF)$. But since $\angle EBF$ and $\angle CBA$ are supplementary (their angle sum plus the angle sum of the two right angles adjacent to them is 360°), $\sin\angle EBF = \sin\angle CBA$. Therefore, the area of $\angle EBF$ is $\frac{1}{2}ac(\sin\angle CBA)$, which is the area of △ABC. Note that this argument did not depend on △ABC being a right triangle.

Fig. **3.6.**

(*a*) The area of △*ABC* is
$\frac{1}{2}a\left(b\sin\left(\theta\right)\right)$; (*b*) the areas of △*ABC* and
△*ACD* are equal when ω and θ are
supplementary and have congruent sides *a*
and *b* in the triangles.

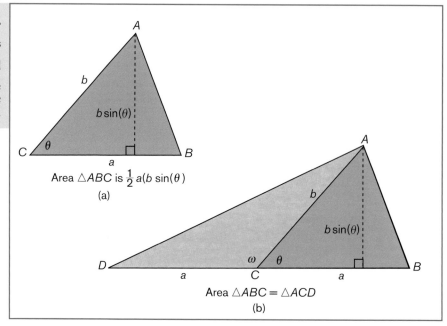

Area △*ABC* is $\frac{1}{2}a(b\sin(\theta)$

(a)

Area △*ABC* = △*ACD*

(b)

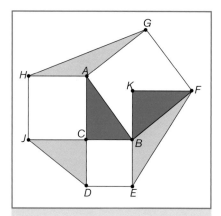

Fig. **3.7.**

A Pythagorean hexagon with △*ABC*
rotated 90° to △*FKB*

If your students are not familiar with trigonometry, you can give them a variety of powerful hints. One of these is pictured in figure 3.4. This figure shows an outcome of the activity Savvy Surveyor. It illustrates the fact that if two sides of a triangle are, respectively, congruent to two sides of another triangle and the included angles are supplementary, then the areas of the triangles are equal, since they can be re-formed as one triangle, with their shared side as a median of the new, single triangle.

Armed with this seemingly simple restatement of the fact that a median divides a triangle into equal areas, the students can deduce that in a Pythagorean hexagon such as that shown in figure 3.5, since ∠*EBF* and ∠*CBA* are supplementary with *CB* = *EB* and *BA* = *BF*, △*ABC* and △*EBF* can be joined, as in figure 3.4, to form a single triangle divided in half by a median. (To make the single triangle, the students could rotate △*ABC* 90° about *B* in a clockwise direction, as in figure 3.7, and then note that $\overline{FB}$ is a median of △*KEF*.) Hence, △*EBF* has the same area as △*ABC*.

Similarly, the students can prove that the other triangles in the hexagon have the same area as △*ABC*. This example reveals how reflecting after completing a proof, as students are encouraged to do in step 4 of Savvy Surveyor, can result in simple new tools for solving seemingly unrelated problems.

Even though the hint given by figure 3.4 greatly simplifies the task posed in step 4 of A Hex on Pythagoras, much remains for the students to do. A primary goal of the activity is to have students analyze their proofs to see if they really need the assumption that △*ABC* is a right triangle. Steps 5–7 take them through this investigation.

In step 7, the students bring their work to a conclusion by generalizing their arguments to prove that for any triangle *ABC*—not just right triangles—the areas of the three external triangles in a Pythagorean hexagon of △*ABC* are equal to the area of △*ABC*. Students can establish the plausibility of this statement first by using dynamic geometry software (see fig. 3.8). As mentioned earlier, the applet Pythagorean

Hexagon Construction, contained on the CD-ROM that accompanies this book, provides a construction that students who do not have access to dynamic geometry software can manipulate.

Assessment

To assess whether students grasp the main ideas of their proofs—especially if they relied on the hints that the activity provided—pay careful attention to their responses in steps 4 and 5. In step 5, they must analyze each deduction in their proof from step 4 to see if it assumed that $\triangle ABC$ was a right triangle. This might be a good time to have the entire class discuss one or more proofs given by students. Some students might suggest developing a dynamic geometric construction and using it to see if the theorem is true for triangles other than right triangles. Indeed, this is a good use of technology and illustrates how dynamic geometric constructions can be helpful for analyzing and further extending one's results after a theorem is proved.

In steps 6 and 7, the students must explore and prove the more general theorem that for any triangle ABC, the areas of the three external triangles in the Pythagorean hexagon of $\triangle ABC$ are equal to the area of $\triangle ABC$. Refrain from giving the students any hint at this stage. Instead, assess whether or not they have grasped the basic ideas in the activity.

A Hex on Pythagoras illustrates how the cycle of exploration, conjecture, and proof can be extended productively by analyzing the assumptions in a proof to see which ones are absolutely necessary. In some contexts, the cycle can lead to a surprising number of intermediate results and generalizations. The next activity, Centroid of Gravity, provides one such context.

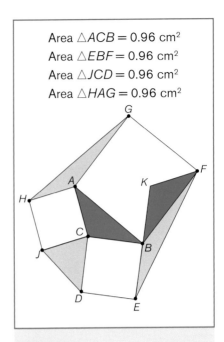

Area $\triangle ACB = 0.96$ cm²
Area $\triangle EBF = 0.96$ cm²
Area $\triangle JCD = 0.96$ cm²
Area $\triangle HAG = 0.96$ cm²

Fig. **3.8.**

Students can use dynamic geometry software to explore the possibility of generalizing to other Pythagorean hexagons.

"In high school, students should be expected to construct relatively complex chains of reasoning and provide mathematical reasons. To help students develop and justify more general conjectures and also to refute conjectures, teachers can ask, 'Does this always work? Sometimes? Never? Why?'" (NCTM 2000, p. 58)

pp. 119–21

If your students do not have access to geometric construction software, you will need to modify the investigations in this activity. By working with grid paper, rulers, and calculators, the students can still discover most of the important ideas. The template "Centimeter Grid Paper" on the CD-ROM enables you to print grid paper for your students' use.

A *locus of points* in the plane is the set of all points in the plane that satisfy a given condition.

The *centroid* of a set of ordered pairs is the point whose coordinates are the averages of the coordinates of all the points in the given set.

Centroid of Gravity

Goals

- Make and test conjectures about the locus of points in the Cartesian plane that have specific distance properties
- Use geometry construction software to explore assumptions, variations, and generalizations of proven geometric relationships
- Use analytical geometry in the proof of conjectured geometric relationships

Materials and Equipment

For each student—

- A copy of the activity sheet "Centroid of Gravity"

For each group of three students—

- Access to a computer with a dynamic geometric construction package (such as Cabri Geometry or the Geometer's Sketchpad)

Discussion

This activity is designed for students who have been introduced to the study of analytic geometry. It investigates sets of points determined by a "squared distance" criterion. The sets of points in the activity always turn out to be circles, and the "squared distance" criterion in a sense generalizes the definition of circles and their centers in the Cartesian plane. This context leads to a surprising number of generalizations that students can explore by using the reasoning and proof cycle.

A *locus of points* in the plane is the set of all points in the plane that satisfy a given condition. One encounters this way of describing geometric figures in the study of the conic sections. For example, a circle is the locus of points P in the plane such that the distance from P to a fixed point C is some specified constant $d > 0$. An ellipse is the locus of points P in the plane such that the sum of the distances from P to fixed points Q and R is some specified constant d greater than the length of $\overline{QR}$.

Centroid of Gravity begins by having the students explore a locus of points whose definition is similar to that of an ellipse. In step 1, they investigate the locus of points P such that the sum of the *squared distances* from P to the fixed points Q and R is the constant $(QR)^2$ (see fig. 3.9a). Using geometric software or grid paper, the students generate many possible locations for P (see fig. 3.9b). Indeed, some students will notice that Q and R both belong to this locus of points. Most students will probably conjecture that the locus of points appears to be a circle with diameter $\overline{QR}$, centered at the midpoint (or centroid) of $\overline{QR}$. Some students will observe that the triangle with vertices P, Q, and R would have to be a right triangle since $(PQ)^2 + (PR)^2 = (QR)^2$. Students who are familiar with traditional theorems about circles can use this observation to prove their conjecture.

Step 2(*a*) asks students to test their conjecture about the locus of points satisfying the given condition. With a dynamic geometric

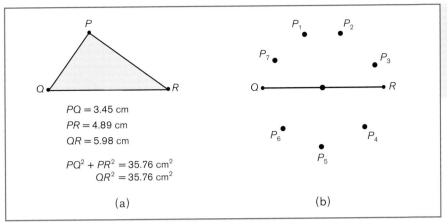

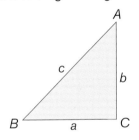

Fig. **3.9.**

Students can use dynamic geometry software to locate different points P satisfying the criterion that $(PQ)^2 + (PR)^2 = (QR)^2$.

construction, the result is very compelling. Students should construct the hypothetical locus, in this case the circle with diameter $\overline{QR}$, and a point P on the locus. As they drag the point P around the circle with diameter $\overline{QR}$ and the distances PQ and PR change, they will see that the sum of the squared distances remains constant at $(QR)^2$. Furthermore, if the students select any point N either outside the circle or inside it, then they will see that $(NQ)^2 + (NR)^2$ is not equal to $(QR)^2$ (see fig. 3.10). They will discover that dragging N to other points in the exterior or interior of the circle does not change this fact.

By the Pythagorean theorem, if $\triangle ABC$ is a right triangle, $a^2 + b^2 = c^2$. The converse is also true: if $a^2 + b^2 = c^2$, then $\triangle ABC$ is a right triangle.

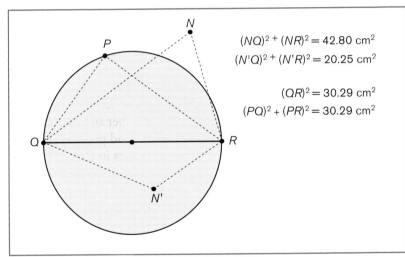

$(NQ)^2 + (NR)^2 = 42.80 \text{ cm}^2$
$(N'Q)^2 + (N'R)^2 = 20.25 \text{ cm}^2$

$(QR)^2 = 30.29 \text{ cm}^2$
$(PQ)^2 + (PR)^2 = 30.29 \text{ cm}^2$

Fig. **3.10.**

A circle is the locus of points satisfying the condition that $(PQ)^2 + (PR)^2 = (QR)^2$.

Step 2(b) calls on the students to prove their conclusions. To prove that the locus of points is indeed the circle with diameter $\overline{QR}$, they must prove that every point P on the circle satisfies the criterion $(PQ)^2 + (PR)^2 = (QR)^2$ and that only those points satisfy the criterion. For $P \neq Q$ and $P \neq R$, this criterion implies, by the converse of the Pythagorean theorem, that $\angle QPR$ must be a right angle. If N is outside the circle, then the secant-secant and secant-tangent theorems guarantee that $\angle QNR < 90°$. If N' is inside the circle, then the chord-chord theorem ensures that $\angle QN'R > 90°$. Either way, neither N nor N' is in the locus of points, since these angles are not right angles. If P is on the circle, the inscribed angle theorem guarantees that except for $P = Q$ and $P = R$, $\angle QPR$ is a right angle. Hence, $\triangle QPR$ is a right triangle. Thus, $(PQ)^2 + (PR)^2 = (QR)^2$. If $P = Q$, then $(PQ)^2 + (PR)^2 = 0 + (QR)^2$, or $(QR)^2$. Therefore, Q is in the locus of points. Likewise, R is in the locus of points. Thus, the locus of points is the circle with diameter $\overline{QR}$.

According to the *secant-secant angle theorem*, if secants $\overleftrightarrow{AB}$ and $\overleftrightarrow{CD}$ of a circle intersect at a point P outside a circle, then the measure of $\angle APC$ is one-half the positive difference of the measures of the two arcs intercepted by $\angle APC$.

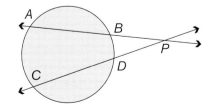

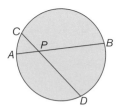

The *chord-chord theorem* holds that if chords $\overline{AB}$ and $\overline{CD}$ of a circle intersect at a point P inside the circle, then $AP \times PB = CP \times PD$.

Unless students are familiar with the angle and segment theorems for circles, they will probably not be able to prove that the locus of points is indeed the circle with diameter $\overline{QR}$. However, a proof is still accessible if the students look at the problem from the perspective of analytic geometry. Step 3 asks the students to create a Cartesian coordinate system in which (–1, 0) represents Q and (1, 0) represents R. This makes QR equal to 2 and $(QR)^2$ equal to 4, and the origin is the midpoint of $\overline{QR}$. Letting (x, y) stand for an arbitrary point P and using the distance formula for points in the Cartesian plane, we discover that the locus criterion $(PQ)^2 + (PR)^2 = (QR)^2$ is true if and only if

$$\left((x + 1)^2 + y^2\right) + \left((x - 1)^2 + y^2\right) = 4.$$

Combining like terms in this equation, we get $x^2 + y^2 = 1$. Thus, a point (x, y) satisfies the locus criterion if and only if (x, y) satisfies $x^2 + y^2 = 1$. That is, (x, y) is on the unit circle, the circle with diameter $\overline{QR}$ (see fig. 3.11).

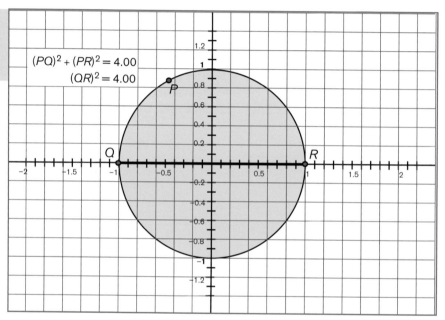

Fig. 3.11.

Using a Cartesian coordinate system to approach the problem

$(PQ)^2 + (PR)^2 = 4.00$
$(QR)^2 = 4.00$

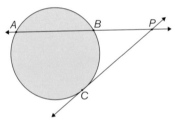

By the *secant-tangent theorem*, if $\overleftrightarrow{PC}$ is tangent to a circle at point C and $\overleftrightarrow{PA}$ is a secant that cuts the circle at points A and B, then $PA \times PB = (PC)^2$.

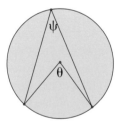

The *inscribed angle theorem* states that the measure of an inscribed angle ψ is one-half the measure of the central angle θ that subtends the same arc. An inscribed angle that subtends a diameter is a right angle, and an angle inscribed on the arc that ψ subtends is supplementary to ψ.

Not only does the algebraic approach to this problem facilitate the reasoning process, it offers students the best tools for exploring variations and generalizations of the result. Step 4 pursues this approach, this time asking students to use it to investigate the locus of points if the criterion changes to $(PQ)^2 + (PR)^2 = K$, where K is any positive constant. Working with dynamic geometric constructions permits the students to verify quickly that the locus is still a circle centered at the midpoint of $\overline{QR}$ for all values of K greater than $\dfrac{(QR)^2}{2}$. The locus is the empty set if K is less than $\dfrac{(QR)^2}{2}$, and it is the midpoint of $\overline{QR}$ if K is equal to $\dfrac{(QR)^2}{2}$.

This persuasive result is difficult to prove without the help of the algebraic representation of the problem. The algebraic proof is straightforward. Using the coordinate representations given in step 3 (see fig. 3.11), $P(x, y)$ is in the locus of points if and only if

$$\left((x + 1)^2 + y^2\right) + \left((x - 1)^2 + y^2\right) = K.$$

Navigating through Reasoning and Proof in Grades 9–12

This expression simplifies to $2(x^2 + y^2) + 2 = K$. Hence, P is in the locus if and only if $x^2 + y^2 = \dfrac{K-2}{2}$. Because $(QR)^2 = 4$, if $K > \dfrac{(QR)^2}{2}$ then $K > 2$ and $\dfrac{K-2}{2}$ is positive, making the locus a circle centered at the origin with radius equal to $\sqrt{\dfrac{K-2}{2}}$. If $K < \dfrac{(QR)^2}{2}$, then $K < 2$ and $\dfrac{K-2}{2}$ is negative, making the locus the empty set. If $K = \dfrac{(QR)^2}{2}$, then $K = 2$ and $\dfrac{K-2}{2} = 0$, making the locus the origin.

Even if your students succeed in proving the conjectures that they generate in this activity, they should be aware that the proofs do not fully explain the situations. Point out that the reasoning and proof cycle always allows for further exploration and deeper explanations. In their work in the previous activity, A Hex on Pythagoras, the students saw how the property that they proved for $\triangle ABC$ had nothing to do with the fact that $\triangle ABC$ was a right triangle. The property turned out to be true for all triangles. Likewise, in Centroid of Gravity, students can discover that the fact that the locus of points turns out to be a circle has nothing to do with the fact that Q and R are the endpoints of a segment. If the students let Q, R, S, and T be the four corners of a square and M be any point in the same plane, they can investigate the locus of points P such that the sum of the squares of the distances from P to the four vertices is the same as the sum of the squares of the distances from M to the four vertices. They will discover that this locus of points is also a circle whose center is the centroid of the square (see fig. 3.12). But even this generalization doesn't exhaust the possibilities, and the result doesn't depend on the points being the vertices of a square!

"In grades 9–12 all students should … use Cartesian coordinates and other coordinate systems, such as navigational, polar, or spherical systems, to analyze geometric situations." (NCTM 2000, p. 308)

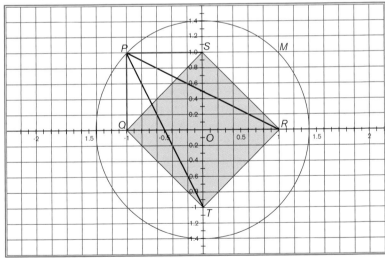

Fig. 3.12.

The circular locus of points P with the sum of the squared distances to the vertices of the square equal to 12, the same sum as for the squared distances from the vertices to point M

Indeed, it is surprising to discover that if W is any finite set of points in the Cartesian plane with centroid $(\bar{x}, \bar{y})$ and C is any circle centered at $(\bar{x}, \bar{y})$ then all the points on C have the same sum of the squared distances to the points in W. Note that $\bar{x}$ is the average of the x-coordinates of the points in W, and $\bar{y}$ is the average of the y-coordinates. In statistics, the point is called the *centroid* of the bivariate data set W. In other words, if M is any point in the plane and d is the sum of the squared distances from M to the points in W, then the locus of points P with the same sum of squared distances is a circle centered at $(\bar{x}, \bar{y})$ and passing through M (see fig. 3.13). The proof of this is provided in the solutions to the activity (see p.171).

(see p.171)

Fig. 3.13.

The circular locus of points whose sum of the squares of the distances to A, B, C, D, E, and F is 126

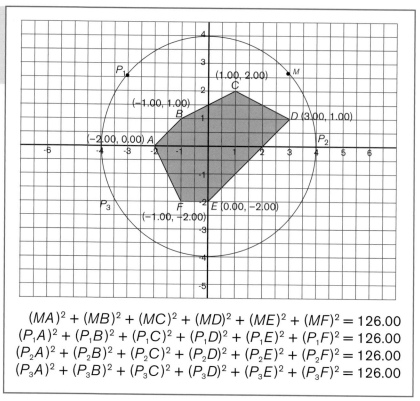

$$(MA)^2 + (MB)^2 + (MC)^2 + (MD)^2 + (ME)^2 + (MF)^2 = 126.00$$
$$(P_1A)^2 + (P_1B)^2 + (P_1C)^2 + (P_1D)^2 + (P_1E)^2 + (P_1F)^2 = 126.00$$
$$(P_2A)^2 + (P_2B)^2 + (P_2C)^2 + (P_2D)^2 + (P_2E)^2 + (P_2F)^2 = 126.00$$
$$(P_3A)^2 + (P_3B)^2 + (P_3C)^2 + (P_3D)^2 + (P_3E)^2 + (P_3F)^2 = 126.00$$

Assessment

There are several fundamental concepts that your students must understand as prerequisites for success with this activity. Perhaps most important, they must understand the equation of a circle centered at the origin with radius r. Review this idea as an introduction to the activity—certainly before your students begin work on step 3. Also, if your students are not familiar with the angle theorems for circles (such as the secant-secant angle theorem), they are likely to have difficulty proving that the locus of points is a circle in step 2. However, not being able to justify their conjecture at this point is acceptable! Reaching an impasse in step 2 can help them appreciate the elegance of the analytic geometry approach in step 3.

When reviewing the formula for the equation of a circle with your students, be sure that they understand that the formula is simply a symbolic representation of the description of the locus of points on the circle. For example, describe the unit circle as "the locus of points P in the plane such that the distance between P and the origin is 1." Show the students the symbolic expression of this idea. Using (x, y) for P and the distance formula

$$D\left[\left(x_1, y_1\right),\left(x_2, y_2\right)\right] = \sqrt{\left(x_1 - x_2\right)^2 + \left(y_1 - y_2\right)^2},$$

the unit circle is the locus of points (x, y) such that

$$\sqrt{\left(x - 0\right)^2 + \left(y - 0\right)^2} = 1.$$

Thus, a point (x, y) is on the unit circle if and only if

$$\sqrt{x^2 + y^2} = 1, \text{ or } x^2 + y^2 = 1.$$

This approach will not only remind students about the equations of circles but also help them make an algebraic statement of the criterion for belonging to the locus of points described in step 1.

Some students may not see the generality of the coordinate geometry proof in step 3. They might suspect that the proof is limited to the case where Q is the point $(-1, 0)$ and R is the point $(1, 0)$. "Even if I've proved the result for these two values of Q and R, don't I still need to prove the result for all values of Q and R?" a student might ask. The answer is no. The students started with two arbitrary points, Q and R, in the plane. The axioms for geometry assure us that there is a coordinate representation of the plane in which Q is $(-1, 0)$ and R is $(1, 0)$. Using that coordinate system, we can prove that the locus is a circle by deriving its equation. Other coordinate systems will work, but the algebra is usually not so simple.

To assess how well your students grasp the main ideas of the activity, have them construct a circle through Q, R, S, and T, the vertices of a square. Ask them to prove that if P is any point on this circle, then the sum of the squared distances from P to the vertices is twice the square of the diameter of the circle. For a proof using algebra, you might suggest productive ways to represent Q, R, S, and T. One such way is to label the points $(-1, 0)$ as Q, $(1, 0)$ as R, $(0, -1)$ as S, and $(0, 1)$ as T on a Cartesian grid. The circle containing these points is the unit circle. Hence, the arbitrary point $P(x, y)$ must satisfy the equation $x^2 + y^2 = 1$.

You can ask your students to prove this result without using algebra, by applying the results of step 2. Here they simply have to note that since the point P is on the circle with diameter $\overline{QR}$ the sum of the squared distances from P to Q and R is the diameter squared. Likewise, since $\overline{ST}$ is also a diameter of that circle, the sum of the squared distances from P to S and T is also the diameter squared. Thus, for any point P on the circle, the sum of the squared distances from P to Q, R, S, and T is twice the diameter squared.

The problems posed in this chapter illustrate the power of reasoning and proof in the context of geometry. The activities also provide examples of the power of geometric contexts to stimulate problem solvers to go through the full cycle of reasoning and proof, which, as the activities have demonstrated, does not always have to end in a proof or a refutation. The next chapter highlights some features of reasoning and proof in the study of data analysis and probability.

"The repertoire of proof techniques that students understand and use should expand through the high school years. For example, they should be able to make direct arguments to establish the validity of a conjecture. Such reasoning has long been at the heart of Euclidean geometry, but it should be used in all content areas."
(NCTM 2000, p. 345)

NAVIGATING *through* REASONING *and* PROOF

Chapter 4
Reasoning and Proof in
Data Analysis and Probability

*"In grades 9–12
students should
gain a deep
understanding of the issues
entailed in drawing
conclusions in light of
variability."*
(NCTM 2000, p. 325)

Statistical and probabilistic reasoning and proof are related to the more traditional forms of reasoning and proof considered so far in this volume. However, they also have their own distinctive characteristics, as this chapter's activities illustrate.

Statistical reasoning begins with data observed in a real-world context and a question arising in that context. It leads to probable conclusions and inferences on the basis of the data. Two key elements in statistical reasoning are *uncertainty* and *variability*, both of which also figure prominently in reasoning about measurement. In spite of the inherent uncertainty of its conclusions, statistical reasoning can produce useful results. Moreover, it permits problem solvers to quantify, with high probability, the uncertainty in their conclusions.

Probabilistic reasoning is a primary component of statistical reasoning and provides the theoretical tools for dealing with uncertainty and variation. Yet, questions about probability, with numerous plausible but uncertain answers, often befuddle students. D'Angelo and West (2000) make the following observation:

> It has been remarked that probability theory is the area of mathematics in which an expert is most likely to blunder. The explanation may be that formulating and solving these questions requires precise language, but the problems often are stated informally and then misinterpreted. We have already seen that the

51

expression "at random" may have more than one interpretation. Ambiguous language can cause difficulties. (p. 174)

In fact, for many students such difficulties come down to a conflict between the basic mathematical meanings of ideas about probability and the less formal, more intuitive uses of them in natural language. Probabilistic questions arise in every situation where natural language permits the question, "What are the chances that...?" Students often encounter difficulties when they need to model such situations in precise and appropriate mathematical terms.

Chapter 4 presents three activities that highlight ways in which reasoning and proof can help students achieve some goals that *Principles and Standards for School Mathematics* (NCTM 2000) outlines for data analysis and probability. These goals include developing students' abilities to—

- use appropriate language and reasoning for modeling probabilistic situations and verify if a model appropriately represents data from the situations;
- design and use simulations to formulate empirical probabilities and probability distributions and compare them to theoretical results;
- use simulations to explore variability in samples drawn from a given population, identify patterns in the variation—whether univariate or bivariate—and use the findings to draw conclusions about the underlying sampling distributions.

Each of the activities in this chapter involves students in the reasoning and proof cycle. Simulations aided by technology allow students to explore probabilistic phenomena and develop conjectures that seem irresistible. Students then use mathematical and statistical reasoning to confirm or refute these empirically based conjectures. Each activity shows the students that determining theoretical probabilities amounts to offering a "proof" of the conjectures that they have generated by inductive reasoning or experimental simulation.

The first activity, Don't Lose Your Heads, illustrates how simulations and statistical reasoning can help clarify some intuitive ideas about probability and the "law of averages." The context is tossing coins. The sample space is finite, consisting of two possible outcomes—"heads" and "tails." These outcomes are *equiprobable*. That is, the coins are assumed to be "fair," so the outcomes are equally likely. The activity illustrates how reasoning and proof can uncover some key features of the *binomial distribution*, which is perhaps the most important distribution for students in grades 9–12 to study and understand.

In the second activity, Derek Jeter Steps into the Box, students prove a result about the *geometric distribution* by using informal reasoning based on simulations, statistics, and averaging. The context involves boxes of cereal containing baseball cards. The question is, if the chance of getting a Derek Jeter card in any single box of cereal is 1 in 4, then how many boxes of cereal should someone expect to purchase to get a Derek Jeter card? This *expected value* problem is complicated by the fact that the underlying sample space has countably infinite outcomes, which are not equiprobable.

The "law of averages" is the intuitive idea that in a large number of independent repetitions of an experiment in which an event A has probability p, the relative frequency of A's occurrence should be close to p. People sometimes mistakenly take the law of averages to mean that the relative frequency of the occurrence of A will "even out" to p in the long run and that a string of luck is thus "bound to change" after sufficient time. This common misunderstanding can lead to the misconception that past outcomes have an impact on future, independent outcomes.

The expected value *is the average value in the long run of a random variable.*

In the final activity, Late Shift at the Orange Bowl, students investigate a problem whose sample space is uncountably infinite. Without assuming any knowledge of the calculus, the activity guides students in using geometric probability arguments based on considerations of length and area to solve the problem. Once again, students use a simulation and data from it, this time to establish an experimental probability that one bus arrives at the Orange Bowl before another bus. Students establish the corresponding theoretical probability by using geometric representations and reasoning.

Students deepen their understanding of statistical and probabilistic descriptions of phenomena when they design their own simulations of these phenomena. To design and implement a simulation requires reasoning, making assumptions, and drawing appropriate conclusions from the results. The simulations used in this chapter are typically quite simple. The required technology is available on all graphing calculators. High school students frequently cannot design simulations on their own without some guidance. Nor do they always know how to analyze the results of the simulations or assess their value. This chapter's discussion of the first two activities supplies hints to offer students as they set up their simulations. The third activity invites the students to design suitable simulations on their own, but even in this case, the students are not left without support. The activity directs them to discuss simulations proposed by their classmates before proceeding, settling on a workable simulation that everyone will use. Two straightforward designs are discussed in the text. All three activities pose questions that guide the students in deriving theoretical results that they can then use to assess the simulations.

Don't Lose Your Heads

Goals

- Perform a simulation to model a binomial experiment and use the results to make a decision about a game of chance
- Use the results of a simulation and basic statistical reasoning
  - Analyze the relationship between *sample size* and *variation* among sample proportions
  - Sort out misconceptions related to the "law of averages"
- Develop the language of probability to describe a binomial experiment and use Pascal's triangle to generate theoretical probabilities for outcomes and events associated with the experiment

Materials and Equipment

For each student—

- A copy of each of the following activity sheets:
  - "Binomial Bonanza"
  - "Data Sheet for 'Binomial Bonanza'"
  - "Pascal's Bonanza"
- A graphing calculator
- A coin to toss

pp. 122–24, 125–27, 128–34

Discussion

Don't Lose Your Heads is an activity in two parts, centering on a game of chance. Part 1, Binomial Bonanza, introduces the scenario for the activity. Brenda has just learned a new game, and she wants her friend Kim to play it with her. Brenda will toss a coin 20 times, and if the percentage of heads is anywhere from 40 percent to 60 percent, then Brenda will win. Otherwise, Kim will win. Several of Brenda and Kim's friends offer opinions on whether Brenda's game offers Kim a fair chance of winning. One friend suggests that the game would be more favorable to Kim's chances if Brenda increased the number of tosses from 20 to 40. Another friend suggests that a game consisting of 10 tosses would be fairer to Kim.

Part 1 guides the students in using simulations, experimental probabilities, and statistical analyses to determine whether Brenda has an advantage in the game. The students record and display their results on a data sheet provided for Binomial Bonanza. Part 2, Pascal's Bonanza, takes the students on an excursion into the theoretical probabilities underlying the game.

Part 1—Binomial Bonanza

Binomial Bonanza begins by asking the students to guess whether tossing the coin 10, 20, or 40 times would be most advantageous to Kim if she decides to play the game with Brenda. In step 2, the students then simulate the 10-, 20-, and 40-toss versions of the game.

To ensure that the students understand how the simulation works, have each student toss a coin five times and record the number and

"By performing simulations with samples of different sizes, students can see that as sample size increases, variation decreases. In this way, they can develop the intuitive underpinnings for understanding confidence intervals."

(NCTM 2000, p. 331)

proportion of heads in their trial of five tosses. In a 5-toss version of the game, Brenda would win if she obtained exactly 2 or 3 heads. Have the students raise their hands if Brenda wins in their simulated trial of the 5-toss game. Count the hands, and ask the students to guess whether Brenda or Kim has the advantage in the 5-toss game. Or do their chances of winning appear to be the same?

Once the students understand the idea of simulating the game, discuss with them how they can use technology to conduct the simulations. Most graphing calculators offer options for generating integers randomly but in equal proportions. For example, share with your students the following two ways of using a TI-83 Plus or TI-84 calculator to perform the simulation of one trial of the 20-toss game:

1. Enter randInt(0,1). This command returns a 0 or a 1 with equal likelihood. If the result is 1, consider the toss to be a head. Press Enter 19 more times, keeping track of the number of 1s that occur, to complete one trial of the experimental 20-toss game. Divide the total number of 1s by 20 to get the proportion of heads that occurred in the trial game.

2. Enter the command randBin(20,.5,1). This command returns the number of heads that occur on 20 tosses of a coin, where the probability of heads is .5. Divide the result by 20 to get the proportion of heads that occurred in the trial game.

Performing the simulations 50 times each for the 10-, 20-, and 40-toss games is time-consuming. The students can reduce the time in various ways, depending on the type of calculator that they are using. For example, on the TI-83 Plus or TI-84 calculator, the command **randBin(40,.5,50)→L1** quickly performs a simulation of the 40-toss game 50 times and stores the 50 results in the list **L1**. This list shows the number of heads that occurred in each of the 50 trials of the 40-toss game. Defining list **L2** as **L2=L1÷40** creates a corresponding list showing the proportion of heads that occurred in each of the 50 trials. If students do not have calculators with the **randBin** command but do have programmable calculators, they can enter a program similar to the one shown in figure 4.1. This program accomplishes the same result as **randBin(40,.5,50)→L1**.

If students store the results of their simulations in lists on their calculators, then they can readily use the calculators to produce histograms and box plots as step 3 of Binomial Bonanza requires. The calculators can also supply summary statistics such as the mean, median, quartiles, and standard deviation for each list.

When the students compare these results in steps 4–7, they should notice that the spread of the histograms and the interquartile range of the box plots get smaller and converge on .5 as the sample size goes from 10 to 20 to 40 tosses. In other words, the more tosses in a game, the more likely the proportion of heads is to fall between 40 percent and 60 percent.

Step 8 asks the students to assess the claims made by Kim and Brenda's friends. These ideas correspond to misconceptions that students commonly have about probability and the "law of averages." In the minds of many students, the fact that the probability of heads in the toss

PROGRAM: FLIP
:For (I,1,50)
:0→A
:For (J,1,40)
:A+randInt(A)→A
:End
:A→L1(I)
:End

Fig. **4.1.**

A calculator program for performing 50 simulations of the 40-toss game

Bernoulli's theorem supposes that the probability of the occurrence of event A in an experiment is *p* and that *N* independent trials of the experiment are made. The theorem, also called the *law of large numbers*, states that if ε is some arbitrarily small positive constant, then the probability that the relative frequency of A's occurrence is outside of the interval [*p* − ε, *p* + ε] goes to 0 as *N* goes to infinity.

"Both plausible and flawed arguments that are offered by students create an opportunity for discussion. As students move through the grades, they should compare their ideas with others' ideas, which may cause them to modify, consolidate, or strengthen their arguments or reasoning. Classrooms in which students are encouraged to present their thinking and in which everyone contributes by evaluating one another's thinking provide rich environments for learning mathematical reasoning."

(NCTM 2000, p. 58)

of a fair coin is .5 rests on a vague notion that half the tosses in a very long string of tosses will result in heads. Some students believe that the more times someone tosses a coin, the more likely he or she is of getting *exactly* 50 percent heads, as long as the number of tosses is even. This idea is false. In fact, as the number of tosses increases, the likelihood of getting heads in exactly 50 percent of the outcomes decreases, finally going to 0. This and related misconceptions are discussed in the solutions section (see pp. 174–176).

The "law of averages" is an intuitive idea that for a large number of independent repetitions of an experiment in which an event A has probability *p*, the relative frequency of A's occurrence should be close to *p*. Bernoulli's theorem makes this idea precise. According to the theorem, if the probability of the occurrence of event A in an experiment is *p* and *N* independent trials of the experiment are made, then for some arbitrarily small positive constant ε, the probability that the relative frequency of A's occurrence is outside the interval [*p* − ε, *p* + ε] goes to 0 as *N* goes to infinity. In the context of the game in the activity, Bernoulli's theorem implies that the probability that Kim wins approaches 0 as the number of tosses increases.

The simulations of the 10-, 20-, and 40-toss games should persuade the students that Kim's chances are best with the 10-toss game. Even here the data suggest that the game is unfair. In the 10-toss game, the theoretical probability that Brenda wins is approximately .66. This means that in the 50-trial simulations conducted by the students in the class, virtually all students should find that that Brenda wins more often than Kim.

It is also possible that several students will find Kim winning most of the games in the 50-trial simulation of the 10-toss game or will find her winning almost as many games as Brenda. Such cases can provide a good backdrop for a discussion of the need for a proof that the game favors Brenda. Most students will claim that the simulations offer a sufficient reason for Kim to decide not to play the game with Brenda. And this point is well taken! However, the simulations do not explain why there is so much variation between Kim's likelihoods of winning the 10-toss game and the 40-toss game. To explain this phenomenon—and indeed to sort out confusion about the law of averages—the students need to delve deeper into the mathematics of probability. Part 2, Pascal's Bonanza, guides students in a theoretical analysis of the possibilities.

Part 2—Pascal's Bonanza

At the beginning of part 2 of Don't Lose Your Heads, the students learn that Kim decides not to play Brenda's game. Brenda protests Kim's decision, contending that Kim merely had a string of bad luck in her simulations and that in fact the theoretical probability of Kim winning is the same as that of Brenda winning. The students must evaluate this claim.

Step 1 provides a discussion of the language and assumptions in the mathematical model of the coin-toss experiment. New vocabulary and ideas are introduced in a list of the assumptions in the theoretical model:

• When an *experiment* consists of one toss of a coin, the only possible *outcomes* are "the coin comes up heads" or "the coin comes up tails."

The set of all possible outcomes of an experiment is called the *sample space of the experiment*.

- The experiment must result in one and only one of the outcomes in its sample space. In other words, the coin cannot come up both heads and tails at the same time.

- The *probability* of an outcome is a number p with $0 \leq p \leq 1$, and the sum of the probabilities of the outcomes in the sample space equals 1.

- The coin is *fair*, which means that each of the outcomes in the sample space has the same probability. In other words, there is nothing about the coin or the process of tossing the coin that favors one outcome over the other.

The goal of step 1 is to give the students a basis for explaining why the probability of heads in the toss of a fair coin is one-half without resorting to a discussion of long-term relative frequencies. The students need to understand that the model applies to a theoretical "fair coin" and not necessarily to the specific coin that they happen to be tossing. In fact, they should keep in mind the statistician's aphorism "All models are wrong; some models are useful," attributed to George E. P. Box.

In steps 2–5, the students examine the sample spaces for the 2-toss, 3-toss, and 4-toss experiments with a fair coin. Making and inspecting tree diagrams and tables, they discover a systematic way to generate the sample spaces and explain why the outcomes are equally likely. They derive a formula for the number of outcomes in the sample space of the n-toss experiment. Later, in step 6, they group the outcomes in the sample spaces according to the number of heads that occur in each outcome. This work leads to filling in a table whose entries form Pascal's triangle.

The tree diagrams give the students powerful visual representations of the recursive nature of most of the patterns that they encounter in Pascal's Bonanza. These representations can be very useful, since the students might not think of—or have access to—the fundamental counting principle when they are proving that the number of possible outcomes in the n-toss experiment is 2^n. This useful principle states that if an event E_1 can occur in m different ways and for each of these m possible outcomes a second event E_2 can occur in n different ways, then there are $m \times n$ different ways for events E_1 and E_2 to occur.

However, without referring to this rule, the students can still conclude from the tree diagrams that the number of outcomes of the n-toss experiment is twice the number of outcomes of the $(n-1)$-toss experiment, regardless of n. The tree diagrams give a compelling picture of the fact that each of the possible outcomes of the $(n-1)$-toss experiment leads to or produces two different outcomes in the n-toss experiment.

With this insight, the students can reason that since the 1-toss experiment has two possible outcomes, the 2-toss experiment has 2×2, or 2^2, possible outcomes. Likewise, the 3-toss experiment has $2^2 \times 2$, or 2^3, possible outcomes. Although mathematical induction offers an approach to a rigorous proof of the general formula, the students have access to a recursive explanation that also leads straightforwardly to the general formula.

"In high school, students can apply the concepts of probability to predict the likelihood of an event by constructing probability distributions for simple sample spaces. Students should be able to describe sample spaces such as the set of possible outcomes when four coins are tossed." (NCTM 2000, p. 331)

The *fundamental counting principle* is a general rule stating that if an event E_1 can occur in m different ways and for each of these m possible outcomes a second event E_2 can occur in n different ways, then there are $m \times n$ different ways for events E_1 and E_2 to occur.

An *event* is a subset of the outcomes in the sample space of an experiment.

Outcome of First Toss	Outcome of Second Toss: H ($p = .5$)	Outcome of Second Toss: T ($p = .5$)
H ($p = .5$)	HH ($p = .25$)	HT ($p = .25$)
T ($p = .5$)	TH ($p = .25$)	TT ($p = .25$)

Fig. **4.2.**
An area model illustrating the equiprobable outcomes in the 2-toss game

Outcome of First Two Tosses	Outcome of Third Toss: H ($p = .5$)	Outcome of Third Toss: T ($p = .5$)
HH ($p = .25$)	HHH ($p = .125$)	HHT ($p = .125$)
HT ($p = .25$)	HTH ($p = .125$)	HTT ($p = .125$)
TH ($p = .25$)	THH ($p = .125$)	THT ($p = .125$)
TT ($p = .25$)	TTH ($p = .125$)	TTT ($p = .125$)

Fig. **4.3.**
An area model illustrating the equiprobable outcomes in the 3-toss game

The tree diagrams can also suggest a way to generate an explanation of why the probability of each of the 2^n possible outcomes of the n-toss game is equal to $\frac{1}{2^n}$. For example, a tree diagram of the outcomes of the 2-toss game may prompt students to use an area model to demonstrate that each of the outcomes of the 2-toss game has a probability of $\frac{1}{4}$ (see fig. 4.2). Building on this result, they can develop an area model showing that each of the outcomes of the 3-toss game has a probability of $\frac{1}{8}$ (see fig. 4.3). In this recursive manner, they can argue that the probability of each outcome of the n-toss game is one-half the probability of each possible outcome of the $(n - 1)$-toss game.

Step 6 presents a table that categorizes the outcomes of the coin-toss experiments according to their numbers of heads (see fig. 4.4). On the activity sheet, data have already been entered in the rows for the outcomes of the 1-toss, 2-toss, and 3-toss experiments. The students must fill in the row for the 4-toss game, and then they must find a pattern that they could use to categorize the outcomes for higher numbers of tosses quickly and efficiently. In the table in figure 4.4, some cells have been filled in on the basis of the pattern.

If the students do not see the pattern themselves, step 6(c) gives them a hint. They are told that Kim notices that to get three heads in the 5-toss experiment, she must get a head on the fifth toss and exactly two heads in the first four tosses *or* get a tail on the fifth toss and exactly 3 heads on the first four tosses. By inspecting the table in light of this observation, the students will see that in the sample space for the 4-toss experiment, there are 6 outcomes with exactly two heads and 4 outcomes with exactly three heads. Thus, they can reason that there are exactly 6 + 4, or 10, outcomes with exactly three heads in the sample space of the 5-toss experiment. This discovery shows them that to complete a cell in the table, they can simply add the number in the cell above it to the number in the first cell above and to its left. This is the classical pattern that characterizes the generation of Pascal's triangle.

Fig. **4.4.**
The table from step 6 with additional cells filled in on the basis of the pattern

Event: Total Number of Heads

Experiment	0	1	2	3	4	5	6	7	8	9	10
1-toss	1	1	0	0	0	0	0	0	0	0	0
2-toss	1	2	1	0	0	0	0	0	0	0	0
3-toss	1	3	3	1	0	0	0	0	0	0	0
4-toss	1	4	6	4	1	0	0	0	0	0	0
5-toss	1	5		6+4		1	0	0	0	0	0
6-toss	1	6					1	0	0	0	0
7-toss	1	7						1	0	0	0
8-toss	1	8							1	0	0
9-toss	1	9								1	0
10-toss	1	10									1

The hint in step 6(c) does more than suggest the pattern in Pascal's triangle. It suggests *why* the pattern works in the context of the coin-toss experiment. If a coin is tossed N times, then the number of outcomes with exactly r heads is $A + B$, where A is the number of outcomes with r heads in which the Nth toss is a tail and B is the number of outcomes with r heads in which the Nth toss is a head. But this means that A is the same as the number of outcomes with r heads in the first $(n - 1)$ tosses, and B is the same as the number of outcomes with $(r - 1)$ heads in the first $(n - 1)$ tosses. This justifies the pattern that determines all of the entries in the table after the first row.

Once the students have established the connection between the coin-toss outcomes and Pascal's triangle, they can readily complete step 7, which asks them to compute the theoretical probability that Brenda would win the 10-toss game. Step 8 directs the students to interactive software available at the Math Forum Web site for generating the rows of Pascal's triangle electronically. Using information that they obtain from the rows of the triangle corresponding to the 20-toss and 40-toss games, the students can calculate the probability that Brenda would win the 20-toss game ($p = .737$) and the 40-toss game ($p = .846$).

Step 9 brings part 2 to a close by asking the students to compare the theoretical results with their experimental results from their simulations in part 1. The experimental probabilities that the students calculated from their 50-trial simulations are likely to be quite close to the theoretical probabilities in most cases. The students should discuss their comparisons. This discussion will probably generate new questions. Students may wonder, for example, if they could determine how many simulated trials to run to optimize the likelihood that the empirical probabilities in the resulting distribution would be accurate. Thus, the reasoning and proof cycle in data analysis and probability, as in other content strands, leads to new questions and explorations.

Assessment

Many students view the theoretical probabilities assigned to events as long-run relative frequencies that are exact and inescapable. This view reflects some confusion about what it means to say that the probability of getting heads is one-half when someone tosses a fair coin. Students sometimes take this statement to mean that the chance of getting heads *exactly* 50 percent of the time goes up as someone continues to toss a coin.

To demonstrate that this idea runs contrary to what actually occurs, have each student in your class toss a coin 4 times. Ask for a show of hands from students who got exactly two heads. Record the number on the board. Then repeat the experiment, with the students now tossing the coin 50 times. Your students are likely to find that about three times as many of them got exactly 50 percent heads in the 4-toss experiment as in the 50-toss experiment.

Faced with this contradictory evidence, your students might revise their claims to say that the more times someone tosses a coin, the closer he or she will come to getting exactly 50 percent heads. This claim is also false. In the students' experiments with 4 tosses and 50 tosses, the students who got exactly 50 percent heads in 4 tosses are unlikely to have gotten 50 percent heads in 50 tosses. In other words, tossing the

"Because conjectures in some situations are not conducive to direct means of verification, students should also have some experience with indirect proofs. And since iterative and recursive methods are increasingly common …, students should learn that certain types of results are proved using the technique of mathematical induction." (NCTM 2000, p. 345)

Math Forum offers interactive software that enables users to generate the rows of Pascal's triangle up to a row whose number they specify. See http://mathforum.org/dr.cgi/pascal.cgi.

coin 50 times instead of 4 times is unlikely to have gotten them closer to exactly 50 percent heads.

To help your students sort out their ideas about probability and express their intuitions about long-run relative frequencies in unambiguous terms, you should point out the need for a more mathematical analysis of the idea of probability and encourage them to reflect on the results of their simulations. The calculator-based simulations of 50 trials should provide a fairly accurate picture of the inherent variability in the phenomenon of tossing a fair coin. The standard deviations and interquartile ranges of the simulated data shrink significantly as the number of tosses increases. The data from the simulations can help your students dispel some of their misconceptions.

The data also point to an appropriate generalization about long-run relative frequencies—the idea that is captured in Bernoulli's theorem. Brenda wins if the proportion of heads is *close* to 50 percent, where *close* is defined as being within 10 percent of 50 percent. The data suggest that the more times the coin is tossed, the better the chances are that the proportion of heads will be *close* to 50 percent.

One goal of part 2, Pascal's Bonanza, is to help students realize that statements such as, "The probability of getting heads is one-half," are about a mathematical model of tossing a coin. The students need to understand that the value determined to be the theoretical probability depends on assumptions about the equal likelihood of outcomes. Therefore, it is important to discuss and assess their responses to step 1 in Pascal's Bonanza very carefully. After considering the assumptions on which calculations of probabilities necessarily rest, the students are asked to say, *on the basis of the assumptions*, what the probability is of getting heads in one toss of a fair coin. Their responses to this question can reveal a great deal about their understanding of probability.

Theoretically, if A and B are two events, then the probability that both A and B occur, $P(A \text{ and } B)$, is the probability that A occurs times the probability that B occurs, given that A occurs. This probability is expressed symbolically as $(P(A) \times P(B|A)$. The assumption that A and B are independent of each other is equivalent to assuming that $P(B|A) = P(B)$, and therefore, $P(A \text{ and } B) = P(A) \times P(B)$. For example, if A is the event "Tails appears on the first toss" and B is the event "Heads appears on the second toss," and the events A and B are assumed to be independent of each other, then the probability of the outcome TH in the 2-toss experiment is $P(T) \times P(H)$, or $\frac{1}{2} \times \frac{1}{2}$.

Rather than giving your students this definition of *independent events*, you should let them reason from tree diagrams or area models that the outcomes in the sample spaces for the 2-toss, 3-toss, and 4-toss experiments are equiprobable.

A nice way to extend the students' work in the two parts of Don't Lose Your Heads is to ask them to redesign Brenda's game to make it fair. This extension will allow you to assess how well your students have grasped both the simulations and the theoretical investigations. The students can find many possibilities for making the game fair. For example, Brenda can toss the coin once with the rule being that she wins if the toss results in heads and loses if it results in tails. Or Brenda can toss the coin 20 times but declare a win only if she gets between 45 percent and 55 percent heads, inclusive. Brenda's probability of winning this variation of the game is .49655, which is close enough to .5 to call the game fair.

Don't Lose Your Heads provides a simple but powerful context for challenging and developing students' understanding of probability and simulations. The activity demands significant reasoning from data gathered in simulations and careful analyses of the meanings of terms. The sample spaces of the experiments are finite and equiprobable.

In the next activity, Derek Jeter Steps into the Box, the students again work with simulations and theoretical considerations, thus continuing to strengthen their understanding of the cycle of reasoning that takes them through exploration, conjecture, and validation or refutation. This time, however, the sample space is infinite.

Derek Jeter Steps into the Box

Goals

- Apply mathematical reasoning and statistical reasoning to analyze a problem that models the geometric distribution
- Perform a simulation with a graphing calculator or a fair die or spinner to compare empirical results with those obtained analytically
- Generate an informal proof, based on long-run relative frequencies, to establish the expected value of a random variable

Materials and Equipment

For each student—

- A copy of the following activity sheets:
  o "Derek Jeter Steps into the Box"
  o "Data Sheet for 'Derek Jeter Steps into the Box'"
- A graphing calculator
- (Optional) A fair tetrahedral die or a spinner with four equal regions

pp. 135–38, 139–40

Discussion

The scenario for this activity introduces the Rise & Shine Breakfast Company, which has launched a campaign to promote sales of its popular cereal, Sweet & Soggy Squares. Each box of Sweet & Soggy Squares now includes a baseball card featuring one of four successful players: Ken Griffey Jr., Pedro Martinez, Ichiro Suzuki, or Derek Jeter. The company guarantees that it has randomly distributed equal numbers of cards for each player in its boxes of Sweet & Soggy Squares. So how many boxes of Sweet & Soggy Squares should a family of Derek Jeter fans expect to buy, on average, before obtaining a card for their favorite player?

This is the *expected value* problem that the students must solve. All the boxes of the cereal are equally likely to contain the card of a particular player. Therefore, the probability that a random box of the cereal contains a Derek Jeter card is $\frac{1}{4}$.

In Don't Lose Your Heads, a trial of an experiment involving Brenda's game consists of a fixed number of tosses, thus making the sample space of possible outcomes finite. By contrast, in Derek Jeter Steps into the Box, a trial of the cereal-box experiment has an infinite sample space *under the assumption of an unlimited supply of the cereal boxes.* That is, the sample space is infinite if the number of boxes that one might purchase before getting the Derek Jeter card is assumed to be unlimited.

Stated in terms of a random variable X representing the number of boxes purchased by a family before getting a Derek Jeter card, the question that the students must answer is, what is $E(X)$, the expected, or average, value of X? Students in grades 9–12 can provide an informal proof that the expected value is 4 boxes. The informal proof that this activity helps them develop relies on averages, relative frequencies, and the Bernoulli theorem, which the discussion of the previous activity describes (see p. 56).

Step 1 of the activity asks the students to *estimate* the average number of boxes of the cereal that a family would purchase before getting a Derek Jeter card. The cycle of reasoning thus begins with exploration, which leads to a conjecture. In step 2, the students explain the reasoning behind the estimate that they gave in step 1. You can use their explanations as an opportunity to analyze whether they are aware of any assumptions that they are making in their estimates. Step 3 calls on the students to identify these assumptions explicitly.

A common error in reasoning that students make is to conclude that if a family purchased five boxes of the cereal without obtaining a Derek Jeter card, then the probability of obtaining the desired card in the sixth box is *higher* than it was before, since Derek Jeter is now "overdue" to appear. This would not be an error if one assumed that only some small finite number of cereal boxes were in circulation. But in the typical situation given, one can easily imagine tens of thousands of boxes of Sweet & Soggy Squares coming into the market as others go out through sales. Thus, a better model involves an assumption that the supply of the cereal boxes is unlimited, with the probability that a random box contains a Derek Jeter card remaining $\frac{1}{4}$ no matter how many boxes a family purchases.

In step 4, the students must describe a method that they believe would provide an effective check of the accuracy of their estimate. See what they come up with individually, and then let them compare ideas with one another. Do they consider possibilities for simulating the experiment? What do they suggest?

To simulate the purchase of a box of the cereal, students can use a fair tetrahedral die, a spinner divided into four equal regions, or a graphing calculator. Each number appearing on the four faces of a tetrahedral die can represent one of the four players appearing on the baseball cards in boxes of Sweet & Soggy Squares. With a graphing calculator such as the TI-83 Plus, students can use the command **randInt(1,4)** to generate the integers 1 through 4 with equal likelihood.

The simulation suggested in step 5 represents the Derek Jeter card by the number 2. A trial of the experiment consists of generating a list of random integers from 1 to 4 and stopping when a 2 appears. The length of the trial is the number of random numbers generated up to and including the 2 and represents the number of boxes of Sweet & Soggy Squares that a family purchased before getting a Derek Jeter card. The students conduct 30 trials. They can use the data sheet accompanying the activity page, and for each completed trial, they should enter every number that appears in the trial as well as the length of the trial. They should also create a frequency table of their results on the data sheet.

Step 6 asks the students to use their data from the 30 trials to determine the average length of a trial, a result that they then compare

"The probability distribution for a random variable can be used to predict the average value in the long run, or the expected value, of the random variable. The concept of expected value is widely used in decision making.… [Students] should realize that the expected value –

• *does not* provide information about the outcome of an individual trial;

• *does* provide information about the average value of the outcome in the long run, over many trials." (Shaughnessy et al. 2004, p. 49)

"In grades 9–12 all students should …

• *use simulations to explore the variability of sample statistics from a known population and to construct sampling distributions; …*

• *use simulations to construct empirical probability distributions."*

(NCTM 2000, p. 324)

with their estimate in step 1. Next, in step 7, they pool their own data on the lengths of trials in 30 simulated trials with the data gathered by five other students. They make a frequency chart to show the pooled results for 180 simulated trials. In step 8, they generate an estimate of the probability distribution for this experiment on the basis of all six students' results.

To help the students arrive at this estimate, step 8 takes them back through the process that they followed in step 6. In the earlier step they found the average length of a trial in their 30 simulated trials and used it to estimate the average number of boxes of Sweet & Soggy Squares that someone would need to buy to get a Derek Jeter card. When students averaged the lengths of their 30 trials, the result was probably close to 4. More important, the average length of a trial is the ratio

$$\frac{\text{Sum of the lengths of the 30 trials}}{\text{30 trials}}.$$

Students should realize that the numerator—the sum of the lengths of the 30 trials—is equivalent to the total number of boxes purchased in 30 trials, and the denominator—the number of trials—is equivalent to the number of Derek Jeter cards obtained in the 30 trials, since each trial stops once a Derek Jeter card is obtained. Thus, the average trial length in the 30 trials is the reciprocal of the relative frequency of Derek Jeter cards in the total number of boxes purchased:

$$\frac{\text{30 Derek Jeter cards}}{\text{Total number of boxes purchased in the 30 trials}}.$$

Step 8(*d*) helps the students recognize that there is nothing magical about 30 trials in this result. They discover that the result is the same if they examine the data that they pooled from 180 trials in step 7: the average length of a trial in the 180 trials is the reciprocal of the relative frequency of the Derek Jeter card in this number of trials. In general, the average length of a trial in N trials is the reciprocal of the relative frequency of the Derek Jeter card in the N trials. By Bernouilli's theorem, as N goes to infinity, the probability that the relative frequency of Derek Jeter cards gets arbitrarily close to $\frac{1}{4}$ approaches 1. Thus, as N goes to infinity, the probability that the average length of a trial gets arbitrarily close to 4 also approaches 1.

The result established informally in the preceding paragraph can be summarized as a general property of geometric distributions: If the probability of success in a Bernoulli trial (which has two possible outcomes—success or failure) is p and N is a random variable representing the number of trials required to achieve the first success in an experiment of independent repeated trials, then the expected value of N, or $E(N)$, equals $\frac{1}{p}$. Less formally, if the probability of success in any given trial is p, the average number of trials required for the first success is $\frac{1}{p}$. Although the activity does not ask the students to articulate this property, steps 8 and 9 both essentially lead them to its discovery.

In step 10, the students use statistical methods to analyze their pooled data set, examining the range of relative frequencies in the combined set. They use their data and analyses to answer the following questions:

- If someone purchases four boxes of Sweet & Soggy Squares, what is the probability that he or she will get a Derek Jeter card?
- If someone wants to make only one trip to the store, how many boxes should he or she buy to have a 90 percent chance of getting a Derek Jeter card?

Answers to both questions, determined on the basis of sample data from 180 simulated trials, are discussed in the solutions section (pp. 182–83). Encourage your students to compare their results based on probabilities determined from the experimental data with results based on theoretical probabilities.

Finally, like the other activities in this book, the activity closes with a task that invites the students to reflect on their results and extend them, in the process posing new questions and proposing new investigations. Step 10(b) asks the students to calculate the probability that a family gets its first Derek Jeter card in its fourth box of Sweet & Soggy Squares. This event corresponds to a sequence of three failures followed by a success, where the probability of failure is .75 and the probability of success is .25. This probability is $.75 \times .75 \times .75 \times .25$, or $(.75)^3 \times .25$. The pattern for calculating the probability $P(N)$ that the first success comes on the Nth trial is evident:

$$P(N) = (.75)^{N-1} \times .25.$$

As N goes from 1 to infinity, the probabilities, $P(N)$, form a geometric sequence (see fig. 4.5), whose infinite sum is 1. This is why the distribution of these probabilities is called the *geometric distribution*.

Fig. **4.5.**
The probability that someone gets his or her first Derek Jeter card in the Nth box in a trial

Length of trial (N)	1	2	3	4	...	N
$P(N)$	$(.75)^0 \times .25$	$(.75)^1 \times .25$	$(.75)^2 \times .25$	$(.75)^3 \times .25$	...	$(.75)^{N-1} \times .25$

Assessment

It is important to let your students discuss the estimates that they give in step 1 for the average number of boxes of cereal that someone will need to purchase to obtain a Derek Jeter card. Some students will guess correctly that the answer is 4, since there are four different cards that are equally likely to occur in any random purchase. However, you should assess the students' reasoning, at the same time making sure that they are considering the reasoning of others students who make different estimates.

The simple statement that there are four equally likely cards does very little to explain why 4 is the average number of boxes that someone needs to purchase to obtain a Derek Jeter card. It would be very unlikely to change the mind of a student who thinks, "You would have to be awfully unlucky if you needed to buy four or more boxes of the cereal to get a Derek Jeter card so the average must be smaller than four."

One of the goals of reasoning and proof is to provide explanations that make sense and show *why* a certain claim is true and competing

"Understanding how to draw inferences about a population from random samples requires understanding how those samples might be distributed. Such an understanding can be developed with the aid of simulations."
(NCTM 2000, p. 329)

conjectures are false. Having one student state, "There are four equally likely cards," offers another student no very powerful reason to think that four is the average number of boxes that someone needs to purchase to get a Derek Jeter card.

In reflecting on their work in step 10, some students might wonder about the average number of boxes that someone would need to purchase to get a complete set of all four baseball cards. You can assess how well the students have understood the key concepts of the activity by asking, "What strategies and approaches would you use to answer this new question?"

Some students will suggest simulations. Encourage them to design their own. Other students will propose using the results of the activity to answer the question by constructing a theoretical argument. Encourage these students to test their conclusions with a simulation. For example, they might reason that since the average number of boxes needed to get a Derek Jeter card is 4, the average needed to get the entire set of four cards is 16. This is a gross overestimate, and a simulation is likely to raise doubts about it very quickly.

Someone intent on collecting all four cards would get the first card in one purchase. The probability that he or she would get one of the three remaining cards in the next purchase would be $\frac{3}{4}$. Hence, by the property of geometric distributions, the aspiring collector could expect to buy an additional $\frac{1}{\frac{3}{4}}$, or $\frac{4}{3}$, boxes to get one of these three cards.

Once the collector had obtained two cards in the set, his or her probability of getting one of the remaining two cards in the next purchase would be $\frac{1}{2}$. He or she could expect to buy another $\frac{1}{\frac{1}{2}}$ or 2, boxes to get one of these. Having obtained three cards in the set, the collector would have a probability of $\frac{1}{4}$ of getting the final card in the next purchase. So the collector could expect to buy another $\frac{1}{\frac{1}{4}}$ or 4, boxes to get this card. Thus, on average, someone who wanted to collect all four cards could expect to buy $1 + \frac{4}{3} + 2 + 4$, or $8\frac{1}{3}$, boxes of Sweet & Soggy Squares to get all four cards.

A rigorous proof of the results in the last activity is beyond the expectation of students who have not studied infinite series. The informal justification, however, provides a sufficient explanation of why the average number of boxes of cereal that someone would need to buy to get a Derek Jeter card is 4 boxes.

In the next activity, Late Shift at the Orange Bowl, the theoretical sample space is uncountably infinite. In this new situation, the students can prove their results using basic geometry.

Late Shift at the Orange Bowl

Goals

- Design a simulation involving random variables that are continuous
- Use geometric and algebraic ideas to model a probability problem

Materials and Equipment

For each student—

- A copy of the activity sheet "Late Shift at the Orange Bowl"
- A sheet of grid paper (template available on the CD-ROM)
- A graphing calculator

pp. 141–43

You can print grid paper for your students' use in this activity from the template "Centimeter Grid Paper," which appears on the accompanying CD-ROM.

Discussion

The scenario in Late Shift at the Orange Bowl is more complex than those that the students encounter in Don't Lose Your Heads and Derek Jeter Steps into the Box. According to the new scenario, the last bus of the day from the Old Town Line in Miami arrives at the Orange Bowl between 11:00 and 11:20 p.m., and the last bus from the New Town Line arrives between 11:10 and 11:25 p.m. The buses' exact arrival times are randomly and uniformly distributed over their respective time intervals.

The scenario introduces the students to two workers, Tony and Rhonda, both employed on the late shift at the Orange Bowl. Tony always arrives on the last Old Town bus, and Rhonda, on the last New Town bus. The students are told that today is Rhonda's birthday, and Tony wants to surprise her with flowers as she climbs off her bus. The question that the students must answer is, what is the probability that Tony's bus will arrive before Rhonda's? In other words, how likely is Tony to succeed in surprising Rhonda as she steps off her bus?

Step 1 challenges the students to design a simulation of the arrivals of the two buses and describe how to use it to derive an experimental probability to answer the question. The students must also explain any assumptions that their simulations make. Unlike the previous activities in the chapter, this one offers no hints about how to design an appropriate simulation.

Many of your students are likely to break the problem into parts. Some may reason correctly about the parts, but the reasoning of others is likely to be flawed. For example, some may reason that if Tony arrives between 11:00 and 11:10 p.m.—as there is a 50 percent chance that he will—then he will certainly arrive before Rhonda, because her bus has no chance of arriving until 11:10. Likewise, if Rhonda arrives between 11:20 and 11:25 p.m.—as there is a $33\frac{1}{3}$ percent chance that she will—then Tony will surely arrive first. Some students may assume that arrival times between 11:10 and 11:20 favor neither Tony nor Rhonda and so simply add their other two probabilities. If they proceed in this way, they will come up with the right answer—$83\frac{1}{3}$ percent—for the wrong reasons. Though the theoretical probability that Tony will arrive before Rhonda happens to be $83\frac{1}{3}$ percent, the students' reasoning is erroneous. On the basis of analyses such as this one, students often design simulations that choose randomly between blocks of time such as 11:00–11:10, 11:10–11:20, or 11:20–11:25.

To show your students the flaw in this strategy, ask them to use it to compute Tony's chances of arriving before Rhonda if her bus arrives between 11:10 and 11:35. Their reasoning will lead them to conclude that Tony's chances in this altered situation are better than 100 percent!

Once the students have designed their own simulations, step 2 asks them to discuss their designs as a class. Guide this discussion closely, since many of the simulations that the students propose are likely to be flawed, as in the example above. The goal of the discussion is to agree on one simulation that everyone in the class will use so that the students can pool their results. Two designs follow for straightforward simulations that provide good models of the arrivals of the buses:

1. The students can think of the arrival times of the buses as integers from 0 to 25, with 0 standing for 11:00 and 25 standing for 11:25. They can then randomly generate an integer T between 0 and 20 to represent Tony's arrival time by using the command **randInt(0,20)** on a TI-83 Plus or TI-84 calculator. Likewise, they can randomly generate an integer R between 10 and 25 to represent Rhonda's arrival time by using the command **randInt(10,25)** on the calculator. They can let each pair of integers (T, R) constitute one trial of the simulation. If $T < R$ in a trial, then the students should count the trial as a success for Tony, since the result indicates that he arrives before Rhonda. Otherwise, they should count it as a failure. On a TI-83 Plus or TI-84 calculator, the command **randInt(0,20)<randInt(10,25)** will generate T and R and display a 1 if $T < R$ and a 0 if $T \geq R$. By using this command and pressing **Enter** 100 times, the students can generate 100 trials of arrivals of the buses. The number of 1s returned is equal to the percentage of times that Tony arrives before Rhonda.

2. The students can think of the arrival times as real numbers between 0 and 25 with a number such as 13.456 representing the moment 11:13.456 p.m. They can randomly generate a real number T between 0 and 20 to represent Tony's arrival time by using the command **20*rand** on a TI-83 Plus or TI-84 calculator (the command **rand** returns a real number between 0 and 1). They can randomly generate a real number R between 10 and 25 to represent Rhonda's arrival time by using the

command **10+15*rand** on the calculator. If $T < R$, they should count the trial as a success for Tony; otherwise, they should count it as a failure. The command **20*rand<10+15*rand** on a TI-83Plus or TI-84 calculator will generate T and R and display a 1 if $T < R$ and a 0 if $T \geq R$. By using this single command and pressing **Enter** 100 times, they can generate 100 trials of arrivals of the buses. The number of 1s returned is equal to the percentage of times that Tony arrives before Rhonda.

The only difference between these two simulations is in the precision of the representations of the arrival times. Some students may argue that the second design simulates the situation more accurately because the use of real numbers permits the treatment of time as a continuum. However, other students may contend that this design counts as successes the situations in which Tony arrives only a fraction of a second ahead of Rhonda and would hardly have time to get off his bus, let alone rush to greet Rhonda as she steps from hers. Take advantage of such differences of opinion to press the students to identify the assumptions in their models. (It turns out that the results of the two simulations are virtually the same, with both yielding successes for Tony in roughly 84 percent of the trials.)

Once your students have agreed on an appropriate simulation, redirect their attention to step 2 on the activity sheet. Now they must use the chosen simulation to gather data. Step 2 calls on each student to perform 100 trials of the simulation independently, recording the number of times that Tony arrives before Rhonda. This percentage is the experimental probability that Tony arrives first.

Step 3 directs the students to pool their data as a class. You can facilitate their work on this task by suggesting that they create a common list containing all the students' experimental probabilities. If there are N students in your class, then the list should contain N experimental probabilities.

In step 3(*b*), the students use the class data to estimate the theoretical probability that Tony arrives before Rhonda. Most students will probably offer the average of the experimental probabilities that the class obtained as the estimate. This is not a bad strategy. But they must also give an estimate of the amount of uncertainty in the value that they report. Their approaches to this estimate are likely to vary. One appropriate strategy for estimating the uncertainty is to multiply the standard deviation of the data set by 2. Thus, if the average of the set of experimental probabilities is .85 and the standard deviation of the set is .02, students can report their estimate of the true probability that Tony arrives first as .85 ± .04. They can use many other methods of estimating, but the interval that they give in their estimate should pick up most of the values in the set of experimental probabilities.

Steps 4 and 5 guide students in an investigation of the theoretical probability that Tony arrives before Rhonda. The students approach the theoretical probability by constructing a geometric representation of the sample space. The activity invites them to think of the arrival times of the two buses in any trial as composing an ordered pair (T, R), and it encourages them to represent the set of all possible outcomes

The students can use the standard deviation that they calculate for their pooled data to estimate the amount of uncertainty in the value that they report for the probability. Multiplying the standard deviation by 2 is a useful rule of thumb, establishing a confidence interval that indicates with high probability that the true value lies within the interval.

Fig. **4.6.**

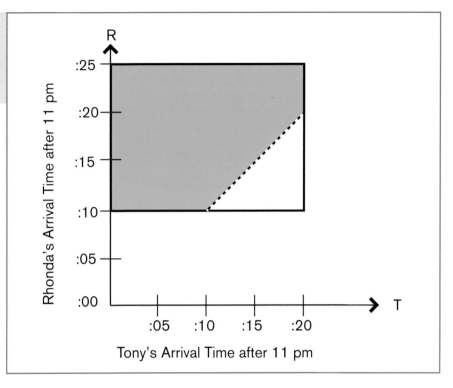

(ordered pairs) as a region. The sample space is the resulting rectangular region shown in figure 4.6.

This representation assumes that the arrival times of each bus constitute a continuum of possibilities. The shaded part of this region corresponds to the event "*T* is less than *R*," or "Tony's bus arrives before Rhonda's." Assuming that the probability of an event is the ratio of its area to the area of the region representing the entire sample space, we can conclude that the theoretical probability of Tony arriving before Rhonda is $\frac{5}{6}$, or $83\frac{1}{3}$ percent. This result should be within the interval estimated by most students on the basis of the class simulation results. In step 6, the students compare or contrast the theoretical result with their estimate.

Step 7 closes the activity by having students pose related probability questions that they can solve by using the geometric methods presented in steps 4 and 5. This task offers the students a chance to continue the reasoning and proof cycle that step 1 began. The following are just two possible questions that the students might ask:

- "What are the chances that Tony and Rhonda will arrive within five minutes of each other?"

- "What are the chances that Tony and Rhonda will both arrive between 11:15 and 11:20?"

Assessment

Some students will find it difficult to design a simulation such as the one that this activity requires. Until they make a comparison of times, they might not see the arrivals of the two buses as independent events, with each bus having its own random arrival time.

While the students are designing their simulations in step 1, take the opportunity to identify students who are having difficulty with the task. Also get a sense of the various simulations that other students are developing. These observations will put you in a good position to facilitate the discussion in step 2. You can steer the discussion in promising directions as well as assess whether the students who had trouble coming up with their own simulations are benefiting from the ideas of their peers.

As the students delve into the area representation of the sample space in steps 4 and 5, carefully assess their conceptions of what probability means in this new context. Most students will say that all the ordered pairs in the region identified as the sample space are equally likely outcomes. This is true in an informal sense. However, because the sample space contains an infinite number of these so-called "equally likely" outcomes, the probability of any one of them occurring is infinitely small—that is, 0. This fact may seem paradoxical to students, most of whom are accustomed to thinking an event with a probability of 0 has no chance of occurring.

The key to resolving these misunderstandings is to have the students analyze what it means to say that the arrival times of a bus are "uniformly and randomly dispersed" throughout the specified interval of its arrival. It is sufficient for the students to interpret this to mean that the probability that the bus arrives in a certain interval of time in the domain is directly proportional to the length of the interval.

Conclusion

Each of the activities in this chapter has made use of simulations as part of the exploration phase of the reasoning and proof cycle. These simulations help students to form appropriate conjectures and make sense of the patterns in the variation that is inherent in probabilistic phenomena. The students use techniques of data analysis and calculations of theoretical probability in validating or refuting the conjectures based on these simulations. The next chapter presents activities in which measurement is the focus in the exploration-conjecture-justification cycle of reasoning and proof.

NAVIGATING *through* REASONING *and* PROOF

Chapter 5
Reasoning and Proof in Measurement

> *"Mathematics should make sense to students; they should see it as reasoned and reasonable. Their experience in school should help them recognize that seeking and finding explanations for the patterns they observe and the procedures they use help them develop deeper understandings of mathematics."*
> *(NCTM 2000, p. 342)*

Giving students opportunities to make sense of mathematics means making them responsible for demonstrating how they understand a concept, for showing their reasoning, and for providing justifications for the work that they do (Hanna 2000). This chapter explores ways of developing students' abilities to reason about and justify ideas of measurement. Students who are able to reason through measurement problems and justify their solutions develop a clearer understanding of the concepts and their connections within mathematics as well as in their daily lives.

The Measurement Standard in *Principles and Standards for School Mathematics* (NCTM 2000) highlights skills and understanding in measurement that students should acquire at the high school level. Students in grades 9–12 should understand which attributes of a given object are measurable as well as how to apply measurement tools, techniques, and formulas appropriately. Students need to have opportunities to explore measurement in various settings, make decisions about the measurement process, and communicate these decisions verbally and in writing.

Principles and Standards suggests that students be provided with opportunities to make comparisons among scales or measurements to determine which are suitable for the object or phenomenon that they are measuring. Your students could make such comparisons in a whole-class setting, or they could explore various objects and ways of quantifying the objects' attributes in small groups. For instance, you could give your students measuring devices such as rulers, measuring cups, and so on, along with several objects to measure, such as Platonic solids, a lump of clay, a bag of beans, and a data set (any set of related numbers). Then you could ask the students to develop various strategies for measuring the objects in ways that describe or predict their important characteristics. In this investigation, your students might consider questions such as the following:

- "What measurements (slant height of a regular tetrahedron, weight of a lump of clay, for example) will help you characterize the object? Explain."
- "How does this measurement relate to other measurements that characterize the object? Does it help you derive or predict other measurements?"
- "How will you carry out the measurement? What tools do you need?"
- "What information will this measurement give to others? What can others learn about the object by using it? With what precision should you report it?"

Students who have opportunities to investigate measurement in this way are more likely to construct a strong understanding of what attributes are measurable and how these measurements contribute to an understanding of mathematical objects. This chapter presents two activities that offer students these kinds of experiences, along with the chance to reason about and justify their work.

The chapter's first activity, Pushing the Perimeter, gives students an opportunity to look more closely at the relationship between area and perimeter in the case of rectangles. Like all the activities in this book, this activity emphasizes the reasoning and proof cycle. The need for justification may not be immediately obvious in the activity. However, it is likely to arise in discussions as the students realize, in the face of conflicting data, that they must support their conjectures and persuade their classmates of the validity of their ideas. Thus, the activity addresses the Reasoning and Proof Standard elaborated in *Principles and Standards* by helping the students develop a sense of what constitutes valid or invalid reasoning.

Optimization problems, including finding the maximum and minimum values of functions, are typical fare in a calculus class. However, students can readily investigate these types of problems at all levels of high school without techniques of differentiation. To find a maximum or minimum value for a given function, students can use calculators, tables, graphs, spreadsheets, and other tools to "narrow in" with increasing precision on an approximate value for the requested maximum or minimum. The chapter's second activity, Maximizing the Measurement, presents two such problems. Part 1, Heavy Metal at Maximum Volume, leads students through an investigation that helps them find a potential upper bound—the maximum volume of a metal box. They make decisions about what intervals to explore to "close in" on the maximum, and they justify their procedures and their final choices of a maximum value. Part 2, Not Only Useful, but Also Orna-Metal, switches from volume to area for the measurement to maximize. This time the students explore the maximum area of a rectangular box top that someone can edge with a narrow strip of trim that is exactly one meter long.

Pushing the Perimeter

Goals

- Decide whether a given conjecture is true or false by conducting a hands-on investigation
- Justify or explain the truth or falsity of the given conjecture
- Look back at a justification or refutation of the given conjecture to identify sources of error

Materials and Equipment

For each student—

- A copy of the activity sheet "Pushing the Perimeter"

For each group of students—

- One or two sheets of grid paper (template available on the CD-ROM) or a small (≈ 12) set of square tiles

Discussion

Students often say that they find true-or-false questions among the most difficult on mathematics tests. The challenge results in part from the fact that students in many mathematics classrooms do not routinely encounter activities that require them to decide whether or not a statement is valid. In fact, students often operate on a simple but risky assumption: "If it's in the textbook, it must be true." Hence, students frequently do not see the need for proof or justification. It is also often the case that mathematics classes reduce proof writing to a mere exercise—just another algorithm to be memorized.

By contrast, Pushing the Perimeter uses a true-or-false question to set the stage for a learning experience that is rich in reasoning and proof. The activity asks the students to investigate a conjecture by a hypothetical student, Kelsey. According to the scenario for the activity, Kelsey's class conducted an investigation of area and perimeter, and then Kelsey made a conjecture: When the perimeter of a rectangle increases, then the area must also increase. The students must decide whether Kelsey's claim is true or false and explain or justify their conclusion.

That the conjecture is false can be demonstrated in a variety of ways. For example, assuming that the length and width of a rectangle are not restricted to whole numbers, the area of the rectangle can remain fixed while its perimeter becomes arbitrarily large. Consider the following argument: If x and y are the dimensions of a rectangle whose area is 100 square units, then $x \times y = 100$ square units. This expression can be rewritten as $x = \dfrac{100}{y}$, since neither x nor y can be zero (otherwise, there would be no rectangle). As the value of y increases (and x decreases), the perimeter will become arbitrarily large while the area remains constant at 100 square units. Thus, the conjecture is false.

However, the conjecture is likely to appear plausible at first glance to many students. This will be evident in their responses to step 1, in

pp. 144–46

You can print grid paper for your students' use in this activity from the template "Centimeter Grid Paper," which appears on the accompanying CD-ROM.

which they make a decision about the conjecture's validity before they begin to investigate it. Once students have convinced themselves one way or the other, they conduct an investigation (steps 2 and 3), reconsider their evaluation of Kelsey's conjecture (step 4), and formulate an argument to convince others of the truth or falsity of the conjecture (step 5). In other words, in step 5 the students should prove the conjecture if it is valid or explain why it cannot be valid, if that is their decision.

Students should know that one counterexample is enough to show that the conjecture is false. The falsity of Kelsey's conjecture is easy to demonstrate with a specific, concrete example, as in the sample of work by a student shown in figure 5.1. The student has drawm a rectangle that is 1 × 124 square units and a square that is 25 × 25 square units. The perimeter of the rectangle is clearly much larger than that of the square:

$$250 \text{ units} > 100 \text{ units}.$$

It is also apparent that the area of the rectangle is much smaller than that of the square:

$$124 \text{ square units} < 625 \text{ square units}.$$

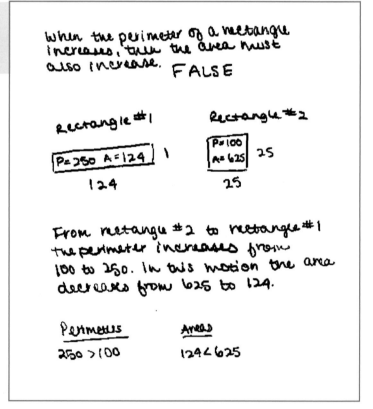

Fig. **5.1.**

Sample of work by a student to demonstrate that Kelsey's conjecture is false

Will a skeptical classmate assert that one such counterexample is simply an exception to the rule? In mathematics, one counterexample is always enough to disprove a statement, but students may transfer ideas from other areas of experience. In these areas, they may be accustomed to encountering numerous exceptions to a rule (such as a rule of English grammar). If a single counterexample meets with skepticism, challenge the students to develop a more generalized counterexample, such as

Navigating through Reasoning and Proof in Grades 9–12

that given earlier for the rectangle with dimensions x and y and an area of 100 square units. Finding a generalized counterexample, or a good explanation, is often the best way to justify a statement in the negative.

You may decide to have your students complete the hands-on investigation of Kelsey's conjecture in pairs. Having a partner will give each student an opportunity for feedback as he or she develops an argument to present to the class. The activity sheet includes a chart for the students to use in recording their findings and poses two "guiding questions" to encourage them to examine a range of rectangular regions:

- "Would considering squares as well as rectangles help you test Kelsey's conjecture?"
- "How could you organize the data in the chart to help you find patterns or trends?"

Reviewing the definition of a square and including squares in the investigation can lead students quickly to the discovery that the original conjecture is false. Some organizations of the data will help the students recognize important patterns, allowing them to discover, for instance, that if one dimension of a rectangle remains constant while the other dimension varies, then the conjecture is true. Organizing their findings may also help the students write a general justification later. Be sure that the students have grid paper or square tiles to help them in their investigation.

Assessment

At first many students will probably claim that the conjecture is true. They may base this claim on data obtained by stretching one dimension of a rectangle while holding the other constant. Under such conditions, the area also must increase. When these students have investigated other possibilities, many of them will probably change their minds. Be attentive to their explanations of their original thinking and their new understanding that Kelsey's conjecture is false. Also pay close attention to the responses of other students who appear not to understand explanations of the falsity of the conjecture.

Sometimes, the way a conjecture is stated can influence a student's reasoning. For example, the statement, "When the perimeter of a rectangle increases, then the area must also increase," can be restated in the following fashion: "If the perimeter of rectangle R is less than the perimeter of rectangle S, then the area of R is less than the area of S." However, some students might not at first view these statements as mathematically equivalent. The restatement of the conjecture explicitly identifies two rectangles and compares them with respect to their perimeters and areas. By contrast, the original statement may seem to the students to refer to one rectangle that they believe maintains its "identity" as its perimeter increases, much as a growing child does: as a child grows taller, the area of his footprint increases.

A student who interprets Kelsey's conjecture in this way might assume incorrectly that the statement, "The perimeter of a rectangle increases," implies that one or both of the rectangle's dimensions increase but does not permit one of its dimensions to decrease while

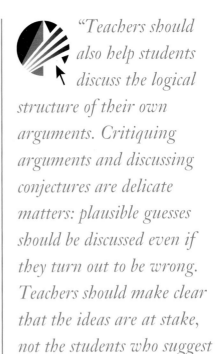

"Teachers should also help students discuss the logical structure of their own arguments. Critiquing arguments and discussing conjectures are delicate matters: plausible guesses should be discussed even if they turn out to be wrong. Teachers should make clear that the ideas are at stake, not the students who suggest them."
(NCTM 2000, p. 346)

the other increases. Working from this assumption, the student might initially disagree with an explanation such as that shown in figure 5.1.

If in the face of such arguments some students remain unconvinced of the falsity of Kelsey's conjecture, you may need to intervene to help them clarify and modify their assumptions. Stress that the perimeter of a rectangle is a value that is determined by two factors—length and width. The student whose work is shown in figure 5.1 has increased the perimeter of the rectangle that is 25×25 square units by stretching one dimension to 124 units while letting the other dimension contract to 1. Kelsey conjectured that when the perimeter of a rectangle increases, the area must also increase. The student's work satisfies Kelsey's "if" ("when") clause, by showing an example in which the *perimeter* increases, while demonstrating that Kelsey's "then" clause is false, by showing that the *area* of the rectangle decreases, from 625 square units to 124 square units.

After your students have persuaded one another—perhaps with your help—that Kelsey's conjecture is false, they will be ready to move on to step 6 on the activity sheet. This step asks them whether Kelsey's conjecture would hold for other shapes. In particular, would it hold for regular and irregular polygons in general? Would it hold for circles?

In fact, Kelsey's conjecture would hold for both circles and regular polygons. As the circumference of a circle increases, so does its radius, with an accompanying increase in the area. In the case of regular polygons, the conjecture is also true, since a regular polygon can be inscribed in a circle, and when the perimeter of the polygon increases, so does the apothem, and this increase in turn produces an increase in the area. A common area formula for a regular n-gon is $A = \frac{1}{2} h \times p$, where h is the apothem and p is the perimeter. The area must increase as the perimeter increases as long as the apothem does not decrease. This can be a nice way to discuss the connection between the perimeter of a regular polygon and its apothem.

This step gives you an opportunity to assess the conclusions that students reach in the new situations after discounting Kelsey's conjecture for rectangles. It is important to ask the students to look back over their original decisions about the validity of the conjecture, and if they initially believed that Kelsey's conjecture was true, they should examine the thinking that led them to the wrong conclusion.

You might also ask your students to consider the converse of the statement—that is, "When the area of a rectangle increases, then its perimeter must also increase." The students should see that the converse of Kelsey's conjecture is also false, and they should be able to come up with counterexamples quickly and easily.

Some students may overextend the results of their work in Pushing the Perimeter by claiming that the measures of perimeter and area are completely independent of each other. This is not true. For a given fixed area, there is a smallest possible rectangular perimeter. And for a given fixed perimeter, there is a largest possible rectangular area. How one finds such optimal values is the focus of the next activity, Maximizing the Measurement. This time, instead of using grid paper or square tiles, the students carry out the exploration phase of the reasoning and proof cycle on graphing calculators or electronic spreadsheets.

The *apothem* of a polygon is a line segment that extends from the center of a regular polygon to a side and is perpendicular to the side. The segment h is the apothem of the regular hexagon below.

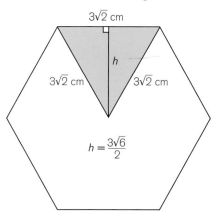

Navigating through Reasoning and Proof in Grades 9–12

Maximizing the Measurement

Goals

- Investigate informal concepts of limit by using successive approximations to estimate maximum and minimum values of volume, area, and perimeter

- Explore different representations of the limiting process

- Use different methods of reasoning to justify chosen representations and resulting optimizations

Materials and Equipment

For each student—

- A copy of each of the following activity sheets:
  - "Heavy Metal at Maximum Volume"
  - "Not Only Useful, but Also Orna-Metal"

For each group of four students—

- A pair of scissors
- A ruler calibrated in centimeters
- A piece of construction paper or tagboard
- Tape
- A graphing calculator or access to a computer with spreadsheet software
- Chart paper or poster board
- Colored markers

pp. 147–50, 151–54

Discussion

Historically, many measurement problems—especially those involving area and volume—were solved by using the limits of successive approximations—methods that eventually became formalized in the calculus. For example, the volume formulas for pyramids, spheres, and many other shapes were derived through the use of such processes long before Newton developed the methods of the calculus. *Principles and Standards* emphasizes the benefits of having students use informal concepts of successive approximation and limits in situations involving measurement.

In Maximizing the Measurement, the students work with shapes whose volumes and areas are easy to calculate—a rectangular prism and a rectangle. They encounter optimization problems under given constraints. Optimization is one of the classical applications of limits and successive approximations that the calculus so elegantly addresses. However, with technology, reasoning, and informal proof, precalculus students can succeed in solving these and other optimization problems.

High school students can use many strategies for investigating an optimization problem. They can calculate values for various measures, find appropriate mathematical expressions for measurable quantities, obtain successive approximations for maximum or minimum values, graph equations, and reason mathematically. All these methods come into play as the students work through the problems in Maximizing the Measurement. To economize on materials and equipment and to

"An important measurement idea, which also helps to establish the groundwork for some fundamental ideas of calculus, is that the measurements of some quantities can be determined by sequences of increasingly accurate approximations."
(NCTM 2000, p. 323)

give your students the benefits of collaboration, assign them to work in groups of four on the activity.

Part 1—Heavy Metal at Maximum Volume

In both parts of the activity, groups of students work together, playing the roles of teams of employees at a sheet-metal company, Blue Metal Designs. In part 1, Blue Metal Designs intends to make an open-topped box that it will market under the name "Heavy Metal." The company will manufacture each box from custom-cut sheet metal, 16 centimeters by 30 centimeters, and will create the box with the largest volume that these dimensions allow.

Each team supposes that management has selected it to come up with the dimensions of the desired box. Team members imagine that they have been given a sample of the sheet metal (tagboard or paper) along with diagrams (see the margin) that show the company's method of making an open-topped box by cutting out four equal-sized squares from the corners of the rectangular sheet metal. The students in each team of four should work together, but each student should also complete his or her own activity sheet.

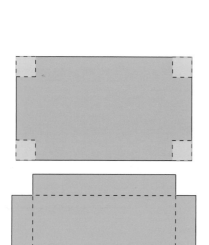

Step 1 asks the student teams for their guesses about the size of the square to be cut from each corner of the sheet metal to create the Heavy Metal box of maximum volume. Then in step 2, working with tagboard or paper, scissors, a centimeter ruler, and tape, each group constructs a full-size model of its box. Next, the group measures the dimensions of its box and computes the box's volume.

Discussing how they performed this task can help the students realize that since they know the dimensions of the sheet metal as well as the size of the square that they are cutting from each corner, they can derive the volume of their box without actually making it and measuring its dimensions. Moreover, comparing the volume of their group's box with those of the boxes constructed by the other groups can help the students appreciate the variability and complexity of the problem.

Steps 3–5 encourage the students to use technology to facilitate their investigations of the Heavy Metal box of maximum volume. Electronic spreadsheets and graphing calculators enable the students to explore the situation by changing the dimensions of the box incrementally. The graphing capabilities of spreadsheet software and graphing calculators also provide visual representations of the mathematical functions under investigation. Thus, the students can make visual sense of the idea of a maximum value.

Steps 4 and 5 guide the students in exploring different methods of "narrowing in" on the data, gradually discovering that the optimal length for the cutout square is $3\frac{1}{3}$ cm. A rigorous proof that this length yields the optimal volume should not be expected of students who have not studied calculus. However, they can validate this solution by using reasoning that is at the heart of optimization in the calculus. When functions are differentiable at their relative maximum and minimum points, their derivatives are 0 at those points. For precalculus students, this translates into an observation that near the optimal point of a graph, the y-values remain nearly the same as the x-values change (see fig. 5.2). Students can use this observation to estimate the maximum error in their approximations of the optimal value. (For example, they can make a scatterplot and estimate an interval visually.)

Part 2—Not Only Useful, but Also Orna-Metal

In part 2, each group of students again acts as a team of employees selected for a design project. The students learn that Blue Metal Designs also sells boxes with rectangular tops. This time, the team is charged with finding the largest rectangular area with a fixed perimeter of 100 centimeters to serve as a box top. Blue Metal Designs will edge this box top with a narrow strip of ornamental trim that is exactly one meter long—precisely long enough to edge a box top with a 100-centimeter perimeter. Metalworkers will fold the trim to form a rectangle, as in the example below, in which the folds occur at the 10-centimeter, 50-centimeter, and 60-centimeter points on a piece of trim aligned with a meter stick.

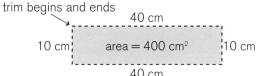

Part 2 guides the team members in applying the methods of exploration and analysis that they used in part 1. As the students set up their tables in step 2, they should quickly realize that if x is the length of the box top in centimeters, then $\frac{100-2x}{2}$, or $50 - x$, is its width in centimeters, and its area is $50x - x^2$ square centimeters (see fig. 5.3). This time, the students can use algebra to find the maximum value, since the graph of the function for the area, $A(x) = 50x - x^2$, is a parabola, and algebra gives tools for finding its vertex.

Some students will know that the x-coordinate of the vertex of the graph of the quadratic function $f(x) = ax^2 + bx + c$ is $\frac{-b}{2a}$. Therefore, they will conclude that the area is maximum when $x = \frac{-50}{-2}$, or 25 centimeters. Other students are likely to use the symmetry of the parabola, noting that the value of the area function is 0 at $x = 0$ and at $x = 50$. Therefore, the x-coordinate of the vertex is exactly halfway between 0 and 50. Hence, the maximum occurs when $x = 25$ cm.

When the length of the rectangle, x, equals 25 centimeters, the width of the rectangle, $50 - x$, also equals 25 centimeters. In general, the rectangle of largest area that can be enclosed by a fixed perimeter is a square.

The students can use this result from their work to validate the optimal solution for the fixed-area problem that concludes part 2. Step 6 presents this problem: A client has ordered a rectangular tray from Blue Metal Designs, specifying only that the area be 900 square centimeters. Management at Blue Metal Designs wants to know the minimum length of trim that the company can use to edge the tray. This time the students must find the smallest perimeter that can enclose a given rectangular area. They work in groups with very little guidance, and each group must make a poster to present its results.

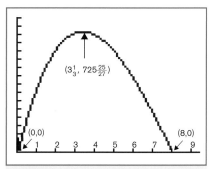

Fig. **5.2.**

Y1(x) = x(16–2x)(30–2x) on a graphing calculator, with the relative maximum indicated (x-axis shows the side length [cm] of the cut-out square; y-axis shows the volume [cm³] of the resulting box)

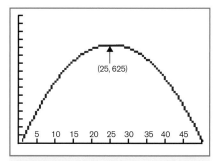

Fig. **5.3.**

Y1(x) = 50x–x² on a graphing calculator, with the vertex of the parabola indicated (x-axis shows the length [cm] of the box top; y-axis shows its area [cm²]

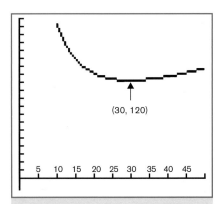

Fig. **5.4.**

Y1(x) = 2x+1800/x on a graphing calculator, with the relative minimum indicated (*x*-axis shows the length [cm] of the rectangular region; *y*-axis shows the area [cm²] of the region)

The fixed-area task reduces to finding the dimensions of a 900 cm² rectangular region with least perimeter. If *x* is the length of the region, then $\frac{900}{x}$ is its width. The perimeter, *p*, is therefore $2x + \frac{1800}{x}$. To minimize *p*, some students will probably use the techniques of successive approximation from part 1, discovering that the limiting value of *x* appears to be 30 centimeters, at which value the rectangle is a square (see fig. 5.4).

Other students will use the results of the fixed-perimeter problem that they have just completed to prove that *p* is minimized when the rectangle's length, *x*, is 30 centimeters, its width, $\frac{900}{x}$ is also 30 centimeters, and the region is a square with a perimeter, *p*, of 120 centimeters. If a rectangle had a smaller perimeter, *p* < 120 cm, then on the basis of the solution to the fixed-perimeter problem, the maximum rectangular area that *p* could enclose would be a square of side $x = \frac{p}{4}$. Therefore, $x < \frac{120}{4}$, or *x* < 30. But such a square would have an area x^2 that would be less than $(30)^2$ square centimeters, which contradicts the fact that the area must be 900 cm².

When a student has a viable candidate for a maximum or minimum solution, such as *x* = 30 cm in the preceding problem, he or she can use a bit of algebra to prove that the candidate is optimal in the interval around it. For instance, on the basis of the graph of $y = 2x + \frac{1800}{x}$ (see fig. 5.4), the student can argue that for some arbitrarily small constant ε, if $x = 30 + \varepsilon$ is positive and yields a smaller perimeter than a length of *x* = 30 cm, then the following would be true: $2(30 + \varepsilon) + \frac{1800}{30 + \varepsilon} < 120$.

The student can simplify this expression as follows:

$$2(30+\varepsilon)+\frac{1800}{30+\varepsilon}<120$$
$$2(30+\varepsilon)(30+\varepsilon)+1800<120(30+\varepsilon)$$
$$(30+\varepsilon)^2+900<60(30+\varepsilon)$$
$$1800+60\varepsilon+\varepsilon^2<1800+60\varepsilon$$
$$\varepsilon^2<0$$

No such ε, positive or negative, exists. Therefore, 30 cm is the optimal side length, and the square is the optimal rectangle for enclosing 900 cm².

Assessment

As your students work on part 1, encourage them to explore possible ways of using more than one type of representation (table, graph, model, and so on) to find and justify their estimate of the maximum volume. Assess the students' understanding of their methods and results by asking questions such as the following:

- "Why is a smooth (continuous) graph an appropriate representation of the situation?"
- "How does the graph represent the maximum value for the side length of the cut-out square [*or* the volume of the Heavy Metal box, *or* the area of the box top to be edged with trim, *or* the distance from one end of a one-meter strip of trim to the point at which the trim must be folded to edge the first corner of the box top]?"
- "What mathematical tools do you have to help you identify the point of maximum value shown on the graph?"
- "Why might a series of tables with successively smaller increments be more helpful than the graph for finding the maximum value?"
- "How do you know that your candidate for the maximum volume is the only possibility?"

In part 2, even though the graphs and the tables make the square appear to be an irresistible solution to both the fixed-perimeter and the fixed-area problem, it is important to challenge your students to try to prove their conjectures. Whether they are zooming in on a point of a graph or refining the increments in a table, they must realize that their approach to the problem is limited by the precision of the tools that they are using and the fact that they can never be certain that they have reached the absolute minimum or maximum. To be sure that their answers are valid—and to persuade others of their validity—they must offer more justification.

You can use your students' work on the final problem to assess their understanding of the activity's main ideas. Have two or three groups with different approaches to the problem present their posters to the entire class. Let students in the class raise questions and answer them during the presentations. Make sure that your students address such questions as, "How do you know that the limit on x, the side length of the tray, is 30 centimeters as you narrow in on the value of x that gives the optimal perimeter?" Or, "Why can't the optimal perimeter be 110 centimeters?"

"In high school, [students'] standards for accepting explanations should become more stringent, and they should develop a repertoire of increasingly sophisticated methods of reasoning and proof."
(NCTM 2000, p. 342)

Conclusion

In Maximizing the Measurement, as in Pushing the Perimeter, students apply their understanding of measurements of length, area, and volume in working with simple geometric shapes. They have had years of experience with the formulas that they use in these activities for perimeter, area, and volume. Furthermore, in these activities, the students work with familiar standard units as they use reasoning and proof to explain patterns in the relationships among these measures. This application of their understanding of the ideas and techniques of measurement to reasoning and justifying conclusions in the area of measurement will deepen their understanding of the need for, and methods of, proof in mathematics in general.

NAVIGATIONS SERIES

GRADES 9–12

NAVIGATING *through* REASONING *and* PROOF

Looking Back and Looking Ahead

Reasoning and proof are at the heart of mathematics. They are the habits of mind of mathematicians. Mathematics doesn't make sense without them. Students gradually learn the norms of logical reasoning and mathematical justification in their school experiences from prekindergarten through grade 12. *Principles and Standards for School Mathematics* (NCTM 2000) expects that all students, by the end of their high school years, will "develop enough confidence in their reasoning abilities to question others' mathematical arguments as well as their own" and will learn to "rely more on logic than on external authority to determine the soundness of mathematical arguments" (p. 346). Indeed, "high school students should be able to present mathematical arguments in written forms that would be acceptable to professional mathematicians" (p. 58).

Attaining these desirable goals is an extremely challenging task. It is no small accomplishment, for example, when students learn to differentiate everyday connotations of words from their more precise, restrictive denotations in mathematical discourse. For example, many students might agree that the statement, "Some Saturdays are rainy," implies that some Saturdays are not rainy. This conclusion interprets the words *some* and *implies* in ways that are customary in everyday speech but inconsistent with conventional mathematical usage. In mathematics, the fact that some Saturdays are rainy tells only that: some Saturdays—at least one, that is—are rainy. The statement *implies* nothing about other Saturdays —indeed, it doesn't even guarantee that there *are* other Saturdays. This example highlights the wide gap that can separate the meaning of words in natural language contexts from their accepted use in mathematical contexts.

Teachers work constantly to move students from the rather loosely defined norms of natural language and context-based reasoning to the more precise, abstract, and exacting norms of mathematical language and reasoning. Following the lead of *Principles and Standards*, the activities in this book suggest an effective avenue for achieving this end—by engaging students in the cycle of reasoning and proof, in rich settings in every area of mathematics.

The exploration and conjecture phases of this cycle offer students opportunities to sort out the meanings of the terms that they are using and familiarize themselves with the contexts in which they must reason. Observing students as they work in these phases can give teachers valuable information about the ways in which their students use words to make sense of problems, investigate them, and reason about them.

For example, in Binomial Bonanza—part 1 of Don't Lose Your Heads—students grapple with ideas and arguments that rest on vague natural-language meanings of probability. They sort through different intuitive ideas about the likelihood of winning a coin-toss game while they also examine experimental results from simulations of the game. These explorations prepare them to make sense of the more precise mathematical ideas of probability that they encounter in Pascal's Bonanza—part 2 of the same activity. Here the students use mathematical reasoning to validate or refute claims made earlier.

The exploration phase of the reasoning and proof cycle lets students develop a meaningful familiarity with a mathematical domain, and the conjecturing phase allows them to recognize and speculate about robust patterns in that domain, often developing compelling theories. However, no matter how persuasive these speculations are, they are not by themselves sufficient as explanations of those patterns. The students must understand that validation—or refutation—is the phase of the reasoning and proof cycle that leads them to a deeper understanding of the mathematics with which they are working. It is this phase of the cycle that brings in logic to connect and confirm—or deny—the generality of the patterns that the students have observed and used as the basis for conjectures.

Every activity in this book, to one extent or another, engages students in validating or refuting their conjectures. Sometimes the point for the students to understand is that they cannot draw a particular conclusion, no matter how strong the evidence may have seemed to be. Such is the situation in the activity Regions of a Circle, where an examination of the first five terms of a sequence seems to reveal a doubling pattern, which abruptly breaks down in the sixth term. At other times, the point for the students to discover is that a pattern that they observed rather casually is actually part of a more far-reaching phenomenon, as in the activity Centroid of Gravity.

In most cases, the goal of the activities in this book is to let the students know that no problem is ever completely explained by a proof or refutation. As Polya (1945) emphasized, looking back at our assumptions and methods leads to new conjectures. Energized by new questions, new predictions, new speculations, we return to the exploration phase of the reasoning and proof cycle.

Technologies that facilitate exploration have transformed classroom activities in mathematics and continue to expand the possibilities. As we look into the future, we see students having opportunities to master far more mathematics than students in the past—as long as we nurture their abilities to reason about and prove results. Research in mathematics education can help establish guidelines and benchmarks for developing these indispensable abilities in the different content strands of school mathematics. But to have an impact, such research must reach teachers in their practice. Finding the means of putting research into action in the classroom is one of the great challenges that we see as we look ahead.

In some sense, *Principles and Standards for School Mathematics* represents a bold experiment in curriculum design—a large-scale exploration in its own right. Between the lines, as well as explicitly, *Principles and Standards* makes conjectures about how best to achieve the vision that it has for all children. As we look ahead, we see the coming decade as a time of validation—or, indeed, refutation—for many of these conjectures—particularly as they relate to the Reasoning and Proof Standard.

We hope that the activities in this book will inspire the creation and implementation of similar activities in classrooms across the curriculum. Thus, whether the conjectures of *Principles and Standards* are ultimately validated or refuted, we may be able to learn enough to enlarge our own goals and refine our own conjectures about how to teach the essential, intertwined topics of reasoning and proof in mathematics.

NAVIGATIONS
SERIES

GRADES 9–12

NAVIGATING
through
REASONING
and PROOF

Appendix

Blackline Masters and Solutions

Probing Products

Name_____

1. Use paper and pencil or a calculator to obtain the products of ten pairs of consecutive whole numbers. List your results below.

   a. What pattern(s) do you notice in the results?

   b. Check several other randomly chosen examples until you are confident of the truth of the pattern(s) that you observed. Write a conjecture about all products of consecutive whole numbers. Write your conjecture in "if … then" form.

   c. Let j represent the smaller of two consecutive whole numbers and k represent the larger. Can j and k both be even numbers? _____ Odd numbers? _____ Explain.

   d. On the basis of your answers in 1(c), what can you conclude about the consecutive whole numbers j and k?

   e. Use your conclusion in 1(d) to write a convincing argument that the conjecture that you made about the products of consecutive whole numbers will always be true.

Name_____

2. Arrays of dots can represent a product. For example, the arrays shown below can represent 2×3 and 4×5, respectively.

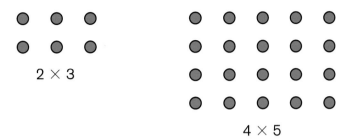

2×3

4×5

Working on a sheet of dot paper, make an array of dots that you can use to provide a visual proof of the conjecture that you made about the product of consecutive whole numbers. Explain your work.

3. So far, you have explored patterns in the products of consecutive whole numbers. Repeat the process from step 1, this time using it to explore the results of squaring an even number (for example, $2^2 = 4$, $8^2 = 64$, and so forth). List ten examples.

a. What generalization(s) can you make about the squares of even numbers?

b. Make a visual argument (using arrays of dots) to justify that the squares of even numbers are always divisible by 4.

c. If n is an even number, then $n = 2 \times k$ for some integer k. (For example, 52 is an even number and $52 = 2 \times 26$.) Use this fact about even numbers to write an argument proving that the squares of even numbers are always divisible by 4.

4. Repeat the process from step 1, using it this time to explore the results of squaring an odd number. List ten examples below.

a. What generalization(s) can you make about the squares of odd numbers?

b. Make a visual argument (using arrays of dots) to justify that one less than the square of an odd number is divisible by 8.

c. If *n* is an odd number, then $n = (2 \times k) + 1$ for some integer *k*. For example, 53 is an odd number and $53 = (2 \times 26) + 1$. Use this fact about odd numbers to write an argument proving that when 1 is subtracted from the square of an odd number, the result is divisible by 8.

Name_____

5. The *triangular numbers* 1, 3, and 6 are shown below.

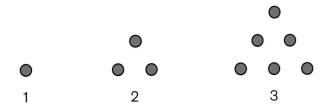

1   2   3

Identify triangular numbers in the arrays representing the squares of odd numbers.

a. Can you relate triangular numbers to your proof that one less than the square of an odd number is divisible by 8?

b. Use a visual representation to justify and explain your reasoning.

Regions of a Circle

Name _____

Consider circles such as those shown below. Non-overlapping regions are formed by connecting *n* points identified on the circles.

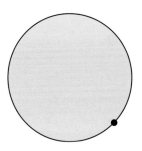
1 point, 1 region

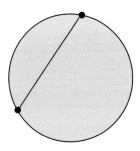
2 points, 2 regions

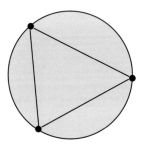
3 points, 4 regions

The first circle, with one point identified, has exactly one non-overlapping region. A *chord* is a segment connecting two points on a circle. Identifying two points on a circle and drawing the chord connecting them creates two non-overlapping regions. Identifying three points on a circle and drawing all possible chords connecting them creates four non-overlapping regions.

Can you find a relationship between the number of points, *n*, identified on a circle and the maximum number of non-overlapping regions that can be formed by all possible chords connecting those points?

1. Explore circles with 4 points identified on them. Draw all possible chords connecting the points. Count the number of non-overlapping regions formed by these chords and the circle. What is the maximum number of non-overlapping regions that you can create with 4 points?

2. What is the maximum number of non-overlapping regions that you can create when you identify 5 points on the circle and draw all possible chords connecting them?

Navigating through Reasoning and Proof in Grades 9–12

Regions of a Circle (continued)

Name _____

3. Summarize your findings so far by filling in the following table for 4 points and 5 points:

Number of points	1	2	3	4	5	6
Maximum number of regions	1	2	4			

4. On the basis of the data in your table for 1–5 points, make a conjecture about the maximum number of non-overlapping regions that are formed by identifying *n* points on a circle and drawing all possible chords connecting them. Write your conjecture as a formula that lets you compute the number of regions from the number of points on the circle.

5. Explain why you think the formula that you gave in step 4 is correct.

6. *a.* What is the maximum number of non-overlapping regions that you can create when you identify 6 points on the circle and draw all possible chords connecting them? *Hint:* You may have to change the position of the points on the circle to ensure that you create the maximum number of interior regions.

   b. Complete the table in step 3 by entering the maximum number of regions created by 6 points.

   c. Describe how you must position the chords to ensure that you maximize the number of non-overlapping regions.

Name _____

d. Do you still think that the formula that you gave in step 4 is correct? _____

Explain why or why not.

7. Sometimes you know the first few terms of a sequence and want to find a formula that gives the other terms. For example, in step 4 you tried to find a formula for the sequence R_n, where R_n stands for the maximum number of regions of a circle determined by placing n points on the circle. You knew from the drawings at the beginning of the activity that $R_1 = 1$, $R_2 = 2$, and $R_3 = 4$.

a. The following table shows the first two terms of a sequence A_n: $A_1 = 3$ and $A_2 = 6$.

n	1	2	3	4	5
A_n	3	6			

Find a formula for A_n that generates these first two terms and use your formula to complete the table.

$A_n =$ _____

b. The function $f(n) = (n - 1)(n - 2)$ is equal to 0 when $n = 1$ or $n = 2$. Add $(n - 1)(n - 2)$ to the expression that you wrote for A_n in 7(a) and verify that the new formula for A_n also gives a valid answer for 7(a). What is your new formula?

$A_n =$ _____

Use your new formula to complete the table:

n	1	2	3	4	5
A_n	3	6			

c. A student named Dania claims that knowing $A_1 = 3$ and $A_2 = 6$ doesn't tell anything about the third term or any other terms of the sequence. "For all we know," Dania says, "A_3 could be the number of students in class today — which I think is 25." After a moment of thought, Dania suggests the formula

Name _____

$A_n = 8(n-1)(n-2) + 3n$. Verify that Dania's formula gives the correct values for A_1 and A_2 and also gives $A_3 = 25$.

8. How many different formulas can be found that fit the first two terms of the sequence A_n?

    _____

    How many different possible values exist for A_3 if all you know are the values of A_1 and A_2?

    _____ Explain your reasoning.

9. Let k stand for any real number in the following formula:
$$P_n = k(n-1)(n-2)(n-3)(n-4)(n-5) + 2^{n-1}.$$

    a. Replace k by any real number and complete the table of values determined by the formula for your value for k.

    $k =$ _____

n	1	2	3	4	5	6
P_n						

    b. Compare the table of values for P_n with the table you completed in problem 3. Do the first five terms of a sequence, such as those in the table from step 3, determine what the 6th term is? _____

    Explain why or why not.

10. What generalization could Dania make about the problem of finding a formula for a sequence when all that anyone knows are the first N terms of the sequence? Explain whether or not you would agree with her, and why.

Algorithmic Aptitude

Name _____

An *algorithm* is a step-by-step procedure for completing a task. Consider an algorithm that gives a quick process for multiplying any two "teener" numbers — whole numbers from 11 through 19. The box below shows how the algorithm for teener multiplication works in the case of 16 × 17.

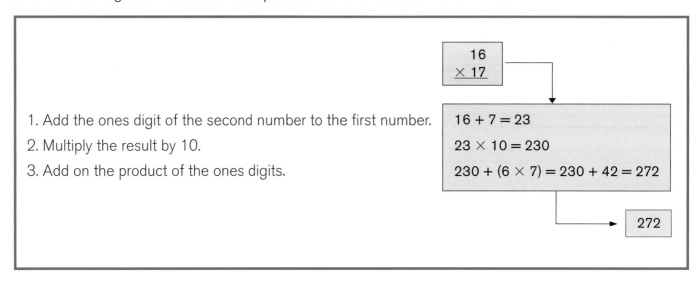

1. Add the ones digit of the second number to the first number.
2. Multiply the result by 10.
3. Add on the product of the ones digits.

$16 + 7 = 23$

$23 \times 10 = 230$

$230 + (6 \times 7) = 230 + 42 = 272$

Does the teener multiplication algorithm hold in every case? Investigate to find out.

1. *a.* Use teener multiplication to find the following products:

$13 \times 19 =$

$16 \times 18 =$

$17 \times 12 =$

Create and test at least two more teener multiplication problems.

Does the algorithm work in every case that you tested?

b. If you can verify that teener multiplication works in the *general case*, you will know that this algorithm always works. Can you use algebra to verify the general case? Every teener number can be expressed in the form $10 + n$, where *n* is a single-digit number from 1 through 9. Suppose that

Navigating through Reasoning and Proof in Grades 9–12

you have two teener numbers, $10 + k$ and $10 + h$, where k and h are both one-digit numbers from 1 through 9. Apply teener multiplication to multiply these two numbers and simplify the result.

c. Use traditional algebraic techniques to multiply $(10 + k)(10 + h)$. Compare the result with your answer in step (b). What can you conclude?

2. We often use algorithms to perform computations, and we trust them to work in all situations. But what if a computational method is flawed? Consider Ted's method of subtraction, for example. In solving the subtraction problems below, Ted sometimes obtains the correct difference, but not always.

$$\begin{array}{cccc} 36 & 85 & 56 & 87 \\ -22 & -37 & -43 & -69 \\ \hline 14 & 52 & 13 & 22 \end{array}$$

a. Identify Ted's incorrect answers and describe his computational method, indicating its flaw(s).

b. Ted's method of subtraction can really mess up a checkbook! If the balance in Ted's checking account is some amount from $10 to $99 and he draws a check for a smaller two-digit amount, what is the largest error that he can record in the transaction? *Hint:* Show several examples of what might happen to various balances that Ted might have in his checkbook, and then make a conjecture.

c. Any two-digit number can be written in expanded place-value form. For example, 38 can be written as $3 \times 10 + 8$, and 24 can be written as $2 \times 10 + 4$. In general, every number from 11 through 99 can be written in the form $a \times 10 + b$, where a is a whole number from 1 through 9, and b is a whole number from 0 through 9. (Zero is included this time since the range of numbers includes 10s: 20, 30, 40, …, 90.)

Let the balance in Ted's checking account be represented by $n \times 10 + r$ and the amount of money that Ted is subtracting be represented by $t \times 10 + s$, where n and t are whole numbers from 1 through 9, and r and s are whole numbers from 0 through 9. Assume that the amount to be subtracted is less than the balance in Ted's account. Using these representations of the two numbers, describe the circumstances in which Ted's method of subtraction results in an error.

d. Use the algebraic forms of Ted's balance and the amount that Ted subtracts to find an algebraic representation of the error that he would make in the circumstances described in 2(*b*).

e. Use your representation of the error in step 2(*d*) to prove your conjecture in step 2(*b*) about the maximum error that Ted can make in using his algorithm.

3. Marie's method for adding one-digit numbers to two-digit numbers is also flawed. Consider her calculations:

$$
\begin{array}{cccc}
67 & 59 & 45 & 36 \\
\underline{+\,9} & \underline{+\,6} & \underline{+\,8} & \underline{+\,4} \\
166 & 125 & 133 & 80
\end{array}
$$

a. Has Marie come up with any correct answers? _____ Describe Marie's computational method, indicating its flaw(s).

b. Like Ted's method of subtraction, Marie's method of addition can also really make a mess of a check-book! If the balance in Marie's checking account is some amount from $10 to $99 and she makes a deposit of a one-digit amount, what is the largest error that she can make in recording the transaction? *Hint:* As in the case of Ted's algorithm, show several examples, and then make a conjecture.

c. Let the balance in Marie's checking account be represented by $a \times 10 + b$ and the amount of money that Marie is adding be represented by c, where a and c are whole numbers from 1 through 9, and b is a whole number from 0 through 9. Using these representations of Marie's addends, describe the error that Marie's method makes.

d. Use the algebraic forms of Marie's balance and the amount that she adds to find an algebraic representation of the error she would make in the circumstances described in 3(*b*).

e. Use your representation of the error in step 3(*d*) to prove your conjecture in 3(*b*) about the maximum error that Marie can make in using her algorithm.

4. Develop a new algorithm for a computation involving two-digit numbers and verify it algebraically. (You might create a new procedure for multiplying a two-digit number by 5, 11, or 12. Or demonstrate a new way to divide a two-digit number by 5. Or invent your own situation!)

Demystifying Magic Squares

Name _____

The square on the right is called a *magic square*. Each small square contains a whole number that is different from all the others. Adding the numbers in any row, column, or diagonal of a magic square always gives the same sum. An $n \times n$ magic square often uses each of the numbers 1 to n^2 exactly once, as in the example of a 4×4 magic square on the right.

1	8	10	15
12	13	3	6
7	2	16	9
14	11	5	4

1. Verify that the 4×4 square shown is "magic" and find its "magic sum."

2. Complete the 3×3 magic square on the right, using each of the numbers 1 through 9 once and only once. What is its magic sum?

	1	
		2

Answer the following questions as you proceed.

a. Explain why 3 cannot occupy the center square. What positions are possible for the 3?

b. Explain why 4 cannot occupy the center square. What positions are possible for the 4?

c. Why can none of the numbers 6, 7, 8, or 9 occupy the center square?

d. By a process of elimination, what number do you conclude must occupy the center square?

Navigating through Reasoning and Proof in Grades 9–12

Name _____

3. *a.* Try to make a magic square with a magic sum of 15 from each square shown below. Existing entries must remain where they appear. If the positions of these entries make it impossible to create a magic square, explain why.

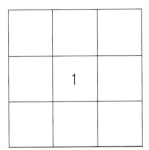

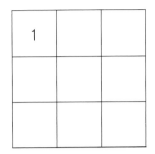

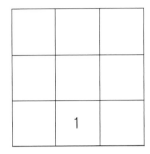

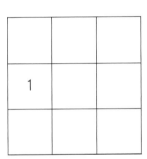

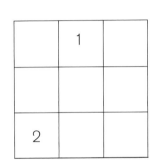

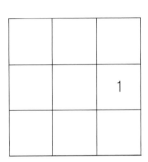

b. What patterns, similarities, or consistent features do you notice among the successful magic squares? List as many as you can.

Demystifying Magic Squares (continued)

Name _____

4 *a*. Entries in an $n \times n$ magic square are not restricted to the numbers from 1 to n^2. Thus, different $n \times n$ magic squares can have different magic sums. For example, 3×3 magic squares can have entries that produce magic sums other than 15. Complete each magic square below by using the given entries, and find the magic sum:

Use the entries 5, 6, 7, 8, …, 13

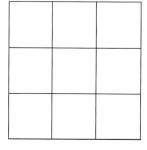

Magic Sum

_____

Use the entries 2, 4, 6, 8, …, 18

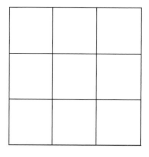

Magic Sum

_____

Select your own entries and create your own magic square below:

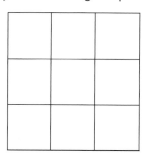

Magic Sum

_____

b. For each example in step 4(*a*), look for a relationship between the sum of all nine entries in the magic square and the square's magic sum. Conjecture how you might determine the magic sum of a 3×3 magic square solely on the basis of its nine entries before placing them in the square.

Demystifying Magic Squares (continued)

Name _____

c. The square on the right uses variables to represent the nine entries in all 3 × 3 magic squares. Use this representation to prove that your conjecture from step 4(b) is true for all 3 × 3 magic squares. *Hint:* If the magic sum is S, then what is true about $(a + b + c) + (d + e + f) + (g + h + i)$, the sum of the sums of the rows?

a	b	c
d	e	f
g	h	i

5 a. In each example in step 4(a), look for a relationship between the average of the nine numbers in the magic square and the number in the center square. Summarize your findings in a conjecture about all 3 × 3 magic squares.

b. Using the representation of the general 3 × 3 magic square in step 4(c), prove that your conjecture from step 5(a) is true. *Hint:* Combine the equations $S = d + e + f$, $S = a + e + i$, $S = b + e + h$, and $S = c + e + g$ to show that e is one-third of the magic sum S.

6. a. Can the set of numbers {1, 3, 5, 7, 8, 10, 12, 14, 16} be arranged to form a 3 × 3 magic square? Justify your answer.

b. Explain why the set of numbers {0, 2, 3, 4, 5, 7, 8, 9, 10} cannot be arranged to form a 3 × 3 magic square.

c. Create your own set of nine distinct whole numbers that cannot be arranged to form a 3 × 3 magic square, and explain why.

Name _____

7. Andy observed that because the center square was always one-third of the magic sum, he always had to pair the four other numbers on the diagonals so that the center number was the average of each pair. The square on the right, which is the same as the one that you completed in step 2, illustrates Andy's discovery. Examine other magic squares to see if you agree with Andy's claim. Justify your conclusions.

$5 + 3$	1	$5 + 1$
3	5	7
$5 - 1$	9	$5 - 3$

8. Andy claims that if he is given any "start-up" whole number a for the center of a 3×3 square and any whole number values for x and y, he can make the square "magic." His procedure begins as shown on the right.

Complete the magic square, expressing the remaining entries in terms of a, x, and y.

$a + x$		$a + y$
	a	
$a - y$		$a - x$

Can you use Andy's method to create a magic square if $a = 8$, $x = 3$, and $y = 2$?

Show your entries in the square on the right.

	8	

9. Does Andy's method of creating 3×3 magic squares always work? Prove or disprove your answer.

10. Do all 3×3 magic squares fit the pattern that Andy has identified? Why, or why not?

3 × 3 Magic Square Grid

Name _____

3 × 3 Magic Square Numbers

Cut out the squares below and use the movable numbers on the activity sheet "3 × 3 Magic Square Grid."

1	8	3
9	2	6
5	6	7

Navigating through Reasoning and Proof in Grades 9–12

Grappling with Growth Rates

Name _____

Sal Snake, CEO of Snake's Oil Corporation, has come to Easybuck to negotiate some oil deals. He must spend a full week in the frontier boomtown. Money comes easy in Easybuck, but rent isn't cheap. Easybuck has only two hotels, both of which make customers pay in advance for the number of nights they will stay. The High Rentz Hotel charges $1,000 per day plus a $5,000 one-time fee for use of an executive meeting suite. At the Rising Sums Hotel, which likes to entice customers in to gamble in the hotel casino but wants to discourage them from hanging out too long, the rent doubles each day from a base rate of $300 per day for a person who also wants to use an executive meeting suite. Sal wants to stay in the same hotel and use the executive suite for seven days. Which hotel should he choose?

1. *a.* Complete the following fee schedules for renting the hotel's room plus executive meeting rooms:

High Rentz	
Length of Stay	*Total Cost*
1 day	$6000
2 days	$7000
3 days	
4 days	
5 days	
6 days	
7 days	

Rising Sums	
Length of Stay	*Total Cost*
1 day	$300
2 days	$600
3 days	$1200
4 days	
5 days	
6 days	
7 days	

b. Write a function that gives the total cost of renting the rooms at the High Rentz Hotel for *n* days.

$H(n) =$ _____

c. Write a function that gives the total cost of renting the rooms at the Rising Sums Hotel for *n* days.

$R(n) =$ _____

d. Use your functions to predict the cost of Sal's stay if he decides to extend his stay to 14 days instead of 7 days.

e. Why are the costs associated with the two functions so different after 14 days?

2. Sam Snake's decision between the High Rentz and the Rising Sums hotels clearly demonstrates that not all mathematical functions are created equal! Some grow faster than others. In many applications, it is important to know just how fast a function grows. You can sometimes analyze the growth rates of a function by looking at the changes in the values for the function as *x* takes on the consecutive integer values 1, 2, 3, 4, 5, and so on.

a. Complete the following tables for the functions $g(x) = 2x + 1$ and $h(x) = 2^x$.

x	$g(x) = 2x + 1$
1	3
2	5
3	7
4	
5	
6	

x	$h(x) = 2^x$
1	2
2	4
3	8
4	
5	
6	

b. The growth rate of a function $f(x)$ at a specific value $x = n$ can be estimated by the size of the difference $f(n + 1) - f(n)$. This difference, which is often negative, is denoted by $\Delta f(n)$ and read as "delta *f* of *n*." It tells the amount by which $f(x)$ changes as *x* increases from *n* to $n + 1$. We define the "growth rate function of $f(x)$" to be the function $\Delta f(x) = f(x + 1) - f(x)$. Use the values from the tables in 2(*a*) to complete the following tables for the growth rate functions of $g(x)$ and $h(x)$.

x	$\Delta g(x)$
1	2 ←This is $g(2) - g(1)$
2	2 ←This is $g(3) - g(2)$
3	
4	
5	

x	$\Delta h(x)$
1	2 ←This is $h(2) - h(1)$
2	4 ←This is $h(3) - h(2)$
3	
4	
5	

Name _____

c. Identify the pattern in each table and find formulas that give the values of $\Delta g(x)$ and $\Delta h(x)$ for any values of x. Prove that your formulas work for all values of x larger than 5.

3. Any function of the form $f(x) = mx + b$, where m and b are constants, is called a *linear function*, and its graph, as you know, is a line. Find the growth rate functions of the following linear functions:

a. $k(x) = 4x - 2$ $\Delta k(x) =$ _____

b. $r(x) = -3x + 2$ $\Delta r(x) =$ _____

4. a. Review your work in step 3. What relationship do you find between $\Delta k(x)$ and the graph of $k(x)$?

     Between $\Delta r(x)$ and the graph of $r(x)$?

b. State and prove a generalization about the growth rate function of any linear function $f(x) = mx + b$.

5. a. Find the growth rate function of the function $d(x) = x^2$.

     $\Delta d(x) =$ _____

b. On a graphing calculator, with graph window set to show $-10 \leq x \leq 10$ and $-20 \leq y \leq 100$, plot the graphs of $d(x)$ and $\Delta d(x)$.

c. Compare the graphs of $\Delta d(x)$ and $d(x)$. For example, when the growth rate function $\Delta d(x)$ is negative, how is the graph of $d(x)$ changing? When $\Delta d(x)$ is positive, how is the graph of $d(x)$ changing?

Grappling with Growth Rates (continued)

Name _____

6. In step 5, $d(x)$ is a *quadratic function*. Its graph is a parabola. Its growth function $\Delta d(x)$ is a linear function. Any function of the form $f(x) = ax^2 + bx + c$, where a, b, and c are constants and $a \neq 0$, is a quadratic function. Use your graphing calculator and follow the process outlined in (a)–(f) below to explore the relationship between quadratic functions and their growth rate functions.

   a. Enter a quadratic function into **Y1** in your calculator's graph editor. Choose your own values for a, b, and c.

   b. Since the growth rate function of $Y_1(x)$ equals $Y_1(x + 1) - Y_1(x)$, enter the expression "**Y1(x+1)−Y1(x)**" into **Y2**.

   c. Graph the functions either separately or together. Find the x value that gives the vertex of the parabola. What do you notice about the graph of $Y_2(x)$ for an x that is close to this x-value?

   d. Make a table of values for both functions starting at $x = 1$ and stepping 1 unit at a time. Be sure to check the values for each function and consider how they are related. Explore the table of values near the vertex of the parabola. Numerically, what happens to the values of the growth rate function near that vertex?

   e. Enter a different quadratic function into **Y1** and repeat steps (b) and (c). Notice that when you change **Y1**, the growth rate function in **Y2** automatically changes as well.

   f. Determine whether or not the growth rate functions corresponding to quadratic functions are always linear. Justify your response.

Grappling with Growth Rates (continued)

Name _____

7. In step 2, you examined the function $h(x) = 2^x$. Any function of the form $f(x) = a(b^x)$ is called an exponential function if a and b are constants, a does not equal 0, and b is positive but not equal to 1. Find the growth rate functions of the following exponential functions and then find the ratio of the growth rate function to the original function. (As in step 6, you can use your calculator to explore the graphs and tables of values of these functions. Just enter the exponential function into **Y1**, enter the expression **"Y1(x+1)−Y1(x)"** into **Y2**, and enter the expression **"Y2(x)/Y1(x)"** into **Y2**.)

a. $t(x) = 2(3^x)$ $\quad \Delta t(x) =$ _____ $\qquad \dfrac{\Delta t(x)}{t(x)} =$ _____

b. $q(x) = -2(5^x)$ $\quad \Delta q(x) =$ _____ $\qquad \dfrac{\Delta q(x)}{q(x)} =$ _____

8. Generalize your results in step 7 by describing the properties of the growth rate functions corresponding to exponential functions. Justify your answers.

9. The following table shows the values of a function $p(x)$ for several values of x. Determine whether $p(x)$ is a linear function, a quadratic function, an exponential function, or none of these. Justify your response.

x	$p(x)$
1	2
2	16
3	54
4	128
5	250

Savvy Surveyor

Name _____

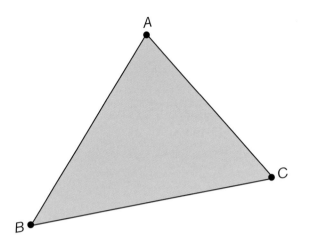

An old, eccentric rancher owns a large region of land in the shape of a triangle. He would like to divide his land evenly between his two adult children. True to his personal preferences, he wants each child to receive a triangular piece of land. Word has gotten around that you are a savvy land surveyor, so he has hired you to come up with a plan to divide the triangular region into two triangular parts of equal area. The old rancher gives you the "map" of his land shown on the right.

1. *a.* Draw a boundary that you think would divide the rancher's land into two triangular regions of equal area. Compare your answer with that of other students in your class.

   b. Can you find common features among the boundaries that you and others have drawn? _____ If so, describe them.

   c. If a boundary does not pass through a vertex of the triangle, can it divide the triangle into two smaller triangles of equal area? _____ Explain your reasoning.

   d. The eccentric rancher wants to be certain that your boundary divides $\triangle ABC$ into two triangular regions of equal area. Prove that your boundary works.

   e. On a separate sheet of paper, construct several different triangles and verify that your method divides each of them into two equal triangular areas. If you have access to dynamic geometric software, create a dynamic construction that shows that the two regions remain equal in area even when you drag a vertex to change the shape of the triangle.

Name _____

2. *a.* The rancher is notoriously indecisive. He tells you that he may decide to keep some land for himself. He asks you to draw an alternate map, showing boundaries that will divide △ABC into three triangles of equal area. Draw appropriate boundaries on the new map of the rancher's land shown on the right. Compare your work with that of other students in your class. How many different solutions did you find altogether?

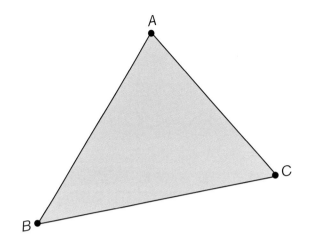

b. The rancher says he likes how the adjacent map looks, but he's not sure that it actually divides the land into three regions of equal area. The point *G* on the map is the *centroid* of △ABC. The centroid of a triangle is the intersection of the triangle's three medians. A median is a segment that connects a vertex of a triangle to the midpoint of the opposite side. If you have access to geometric construction software, use it to demonstrate that the three medians of any triangle intersect at a common point. In any case, explain to the rancher whether or not the three triangles have the same area.

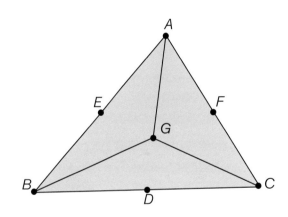

3. *a.* The construction of the centroid of a triangle actually divides the triangle into six triangular regions, as illustrated to the right. If you have access to dynamic geometric software, create a triangle and its centroid as a dynamic construction that you could use to show the rancher that for every new triangle that you create by dragging a vertex, the six regions are always equal in area.

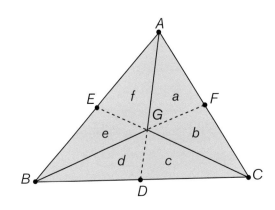

b. If you can prove that the areas *a, b, c, d, e, f* of the six regions created by a triangle's centroid are the same, then you can prove that the three triangles in step 2(*b*) have the same area, since each one is composed of two of the smaller triangles. Prove that the six regions formed by the medians of a triangle are always equal in area. *Hint:* Start by proving that $a = b$, $c = d$, and $e = f$.

4. a. On a separate sheet of paper, trace the triangle that appears at the beginning of the activity and draw in a boundary that divides it into two triangles of equal area. Carefully cut out the two triangles and compare them. Are they congruent?_____ Which features are the same, and which are different?

b. In the following figure, suppose that ∠*ABC* and ∠*DEF* are supplementary, with *AB = DE* and *BC = EF*.

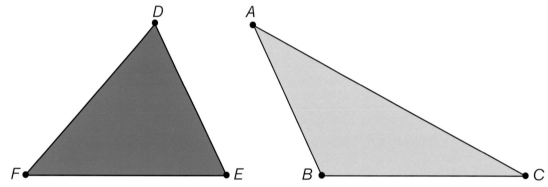

Are the areas of ∠*ABC* and ∠*DEF* therefore equal? _____ Justify your answer.

c. State a generalization about triangles that summarizes what you found in step 4(*b*).

d. Compare your generalization with those of others in your class. Are all the generalizations saying essentially the same thing?

A Hex on Pythagoras

Name _____

The Pythagorean theorem is one of the most famous theorems of geometry. Constructing the exterior squares on the sides of a right triangle, as in the upper figure, helps in visualizing the Pythagorean theorem. The lowercase letters stand for the lengths of the sides of the triangle and squares. If you start with a right triangle and construct the squares on each of its sides, you can connect vertices of the squares, as in the lower figure, to form a hexagon. This activity calls such figures "Pythagorean hexagons," since their exterior boundaries are hexagons formed by connecting vertices of the squares on the sides of triangles. Take the following steps to investigate these hexagons.

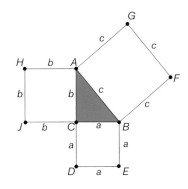

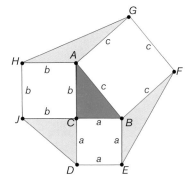

1. Using dynamic geometry software, construct a Pythagorean hexagon, starting with a right triangle. Make sure that your hexagon is what the software refers to as a "construction" rather than a "drawing."

2. Explore the general features of Pythagorean hexagons by measuring different components in the figure and by dragging vertices to change the figure from one hexagon to another. What relationships do you notice? *Hint:* Measure the areas of all the squares and triangles in the figure as well as the lengths of the segments.

3. State several conjectures that you believe are true about all Pythagorean hexagons. Compare your conjectures with those of other students in your class.

4. Suppose that a student makes the following conjecture: "In any Pythagorean hexagon, the areas of the four triangles are equal." Can you prove this conjecture? *Hint:* Try to show that each of the three external triangles has the same area as the right triangle that generates the hexagon.

Name _____

5. Examine your work in step 4. Does your proof make use of the fact that the original triangle was a right triangle? _____ Compare your proof with those of others in your class. To prove that all the triangles in the Pythagorean hexagon are equal in area, must you assume that the original triangle is a right triangle? _____ Explain.

6. Modify your construction of the Pythagorean hexagon so that the original triangle is not a right triangle, as in the figure shown. Use your construction to investigate the properties of Pythagorean hexagons of this type. Explain how the results of your investigation support or refute the conclusion that you reached in step 5.

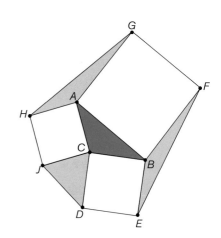

7. Prove that for any triangle △*ABC* — not just a right triangle — that someone uses to generate a Pythagorean hexagon, the areas of the four triangles in the hexagon are equal.

Centroid of Gravity

Name _____

Often mathematicians define a geometric figure as a set of points, or a "locus of points," sharing some specified property. For example, a circle of radius r centered at the origin of the Cartesian plane is defined as the locus of points (x, y) such that the distance of (x, y) to $(0, 0)$ is r. In the figure on the right, a point $P(x, y)$ is on the circle of radius 3 centered at $(0, 0)$ if and only if

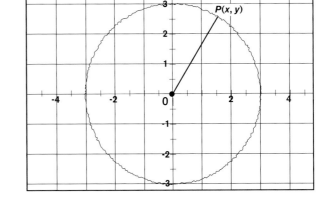

$$\sqrt{(x-0)^2 + (y-0)^2} = 3,$$

or, by squaring both sides, $x^2 + y^2 = 3^2$. In the steps below, use dynamic geometry software or paper and pencil to explore several figures, each of which is a locus of points satisfying a simple property.

1. *a.* Construct a line segment $\overline{QR}$ and measure its length. Call the length d. Find six or more points P with the following property: The square of the distance from P to Q plus the square of the distance from P to R is equal to d^2. The figure on the right illustrates the situation with an example.

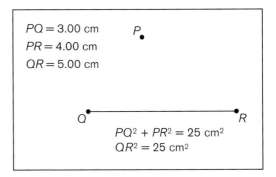

b. Compare the points that you found with those found by other members of your group. Make a conjecture about the geometric figure that consists of all the points with the property given in 1(*a*).

Conjecture:

2. *a.* On the geometry software screen or the sheet of paper where you constructed $\overline{QR}$, construct the figure that you identified in step 1(*b*) as the locus of points P with the property that the sum of the squares of PQ and PR is d^2. Test your conjecture by determining if other points on your figure

Name _____

satisfy the property. With geometry software, you can do this by dragging a point along the figure while updating the measurements and calculations. Does your conjecture appear to be true? _____ If not, modify it and repeat the process. State the conjecture that you ultimately reach.

Conjecture:

b. Prove your conjecture.

3. *a.* On a dynamic geometry software screen with a Cartesian grid, or on the grid to the right, label the point (−1, 0) as Q and the point (1, 0) as R, and construct the figure that you conjectured to be the locus of points in 2(*a*).

b. Find each of the following:

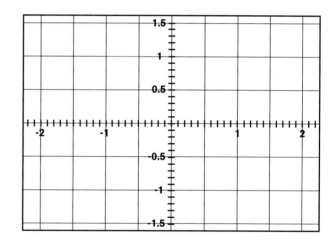

$QR =$ _____

$(QR)^2 =$ _____

Give the equation of the locus of points (x, y) such that the squared distance of (x, y) to (−1, 0) plus the squared distance of (x, y) to (1, 0) is $(QR)^2$.

c. Prove that the figure that you constructed as the locus of points in 3(*a*) is identical to the set of points satisfying the equation that you gave for the locus of points in 3(*b*).

Name _____

4. As demonstrated in step 3, stating a conjecture in terms of Cartesian coordinates and using algebra can sometimes make a proof simpler. An algebraic approach can also make it easier to generalize and extend results that have been proven.

a. Using dynamic geometry software or graph paper, construct a line segment $\overline{QR}$ and investigate the locus of points P such that the squared distance from P to Q plus the squared distance from P to R is some fixed value K that you choose, where K is greater than $(QR)^2$. For each value of K that you might choose, what does the locus appear to be?

b. Using dynamic geometry software or graph paper, investigate the locus of points P such that the squared distance from P to Q plus the squared distance from P to R is a fixed value K, where K is less than $(QR)^2$. For each value of K, what does the locus appear to be?

For what values of K is the locus the empty set?

c. Select a value of K greater than $(QR)^2$ and prove that the locus of points fits the conjecture that you made in 4(*a*).

d. Examine your argument in 4(*c*). How would your proof change if you chose other possible values of K for K greater than or less than $(QR)^2$?

Binomial Bonanza

Name _____

Don't Lose Your Heads — Part 1

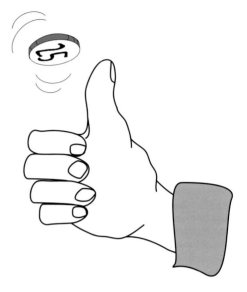

"Let's play a new game I just learned," Brenda urges Kim. "If I win, you treat me to lunch; otherwise, lunch is on me. Here's how it works. I toss a coin 20 times. If 40 to 60 percent of all my tosses turn out heads, I win. Otherwise, you win."

Kim looks skeptical.

Andrea chimes in. "Make Brenda toss the coin 40 times, Kim. It will give you more chances to win."

"No," objects Tracy. "The probability of getting heads is one-half. That means the more times you toss the coin, the better your chances of getting exactly 50 percent heads. Brenda will be more likely to win if you toss it 40 times."

Thia nods. "I have to agree, Kim. Your chances of winning are better if Brenda tosses her coin only 10 times, since the percentage of heads will definitely get closer to 50 percent the more times she tosses it."

"Don't listen to them, Kim," says Brenda. "Look, in 20 tosses, I win only if I get 8, 9, 10, 11, or 12 heads. You have a lot more chances of winning than I do."

Confused by all the advice, Kim says, "Brenda, I would buy you lunch anyway, but you know how competitive I am. So I'm going to experiment a little before I decide whether or not to play your game."

1. Which number of tosses — 10, 20, or 40 — do you think would give Kim her best chance of winning?

Why do you think so?

2. a. Working with a partner, assist Kim in making her decision by performing a simulation. Your teacher will give you directions for simulating games consisting of 10, 20, and 40 tosses. Decide who will run the simulations — you or your partner — and who will record the results. Then run the simulated 10-toss game, 20-toss game, and 40-toss game fifty times apiece, and enter the percentage of heads in each game on the accompanying data sheet for "Binomial Bonanza."

b. Using the grid on the second and third pages of the data sheet, create three histograms to show the fifty percentages that you and your partner have obtained for each of the simulated games.

Binomial Bonanza (continued)

Name _____

3. In the space above the following scale, create three box-and-whiskers plots, one for each version of the game, to summarize the results of the simulations.

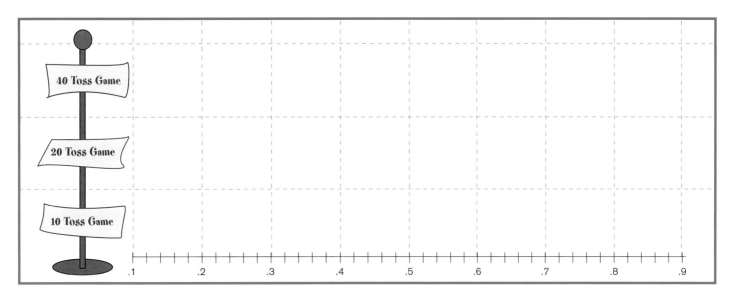

4. On the basis of your data, histograms, and box-and-whiskers plots, which version of the game do you think would give Kim the best chance of winning? _____ Explain why you think so.

5. What trends do the box-and-whiskers plots exhibit as the number of tosses specified for the game increases from 10 to 40?

6. What trends do the histograms exhibit as the number of tosses specified for the game increases from 10 to 40?

7. What conclusions and conjectures can you make from your simulations and data analyses?

Name _____

8. Andrea, Tracy, Thia, and Brenda all make claims about Kim's chances of winning. Assess the validity of each of their claims. On the basis of your results from the simulations, tell why you agree or disagree with each claim.

Andrea's claim:

Tracy's claim:

Thia's claim:

Brenda's claim:

9. Why is the probability of getting heads on a toss of a fair coin said to be one-half? Explain what this means.

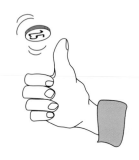

Data Sheet for "Binomial Bonanza"

Name _____

Don't Lose Your Heads – Part 1

10-Toss Game	20-Toss Game	40-Toss Game

Data Sheet for "Binomial Bonanza" (continued)

Name _____

A Histogram of 50 *p*-Values for the 10-Toss Game

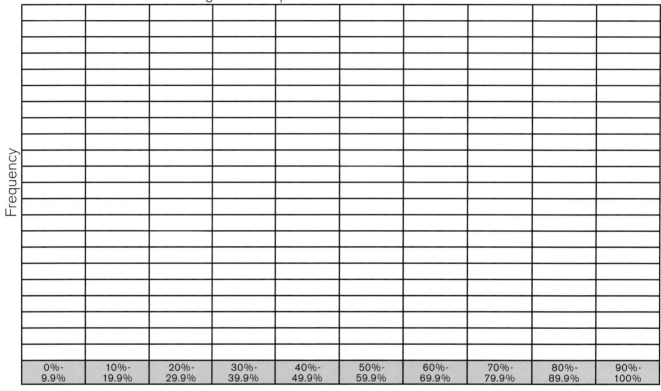

| | 0%-9.9% | 10%-19.9% | 20%-29.9% | 30%-39.9% | 40%-49.9% | 50%-59.9% | 60%-69.9% | 70%-79.9% | 80%-89.9% | 90%-100% |

Frequency

A Histogram of 50 *p*-Values for the 20-Toss Game

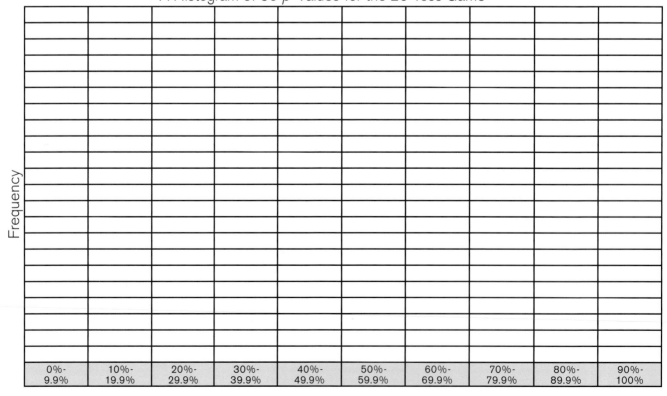

| | 0%-9.9% | 10%-19.9% | 20%-29.9% | 30%-39.9% | 40%-49.9% | 50%-59.9% | 60%-69.9% | 70%-79.9% | 80%-89.9% | 90%-100% |

Frequency

Navigating through Reasoning and Proof in Grades 9–12

Data Sheet for "Binomial Bonanza" (continued)

Name _____

A Histogram of 50 *p*-Values for the 40-Toss Game

Frequency	0%-9.9%	10%-19.9%	20%-29.9%	30%-39.9%	40%-49.9%	50%-59.9%	60%-69.9%	70%-79.9%	80%-89.9%	90%-100%

Pascal's Bonanza

Name _____

Don't Lose Your Heads – Part 2

Kim decides not to play Brenda's game.

"Gee, Kim," protests Brenda. "You think your experiments prove that the odds favor me. But they only prove that you had a streak of bad luck. Your luck is bound to change!"

"Maybe," replies Kim. "But maybe my results weren't just bad luck, Brenda."

"Come on, Kim," says Brenda. "You know what? I really think the probability of winning is theoretically the same for both players."

Kim has a dilemma. She knows that she is basing her decision on the results of experiments. But in the simulations of all three versions of the game, the *relative frequency* of Brenda's winning the experimental games was higher than the relative frequency of Kim's winning. The relative frequency of winning is the total number of wins divided by the total number of games.

To resolve her dilemma, Kim decides to consider the theoretical probabilities of winning – not just the experimental probabilities, or relative frequencies, resulting from the simulations.

1. Help Kim resolve her dilemma. To calculate theoretical probabilities, you must make certain assumptions:

   • When an *experiment* consists of one toss of a coin, the only possible outcomes are "the coin comes up heads" or "the coin comes up tails." The set of all possible outcomes of an experiment is called the *sample space of the experiment*.

   • The experiment must result in one and only one of the outcomes in its sample space. In other words, the coin cannot come up both heads and tails at the same time.

   • The *probability* of an outcome is a number p with $0 \leq p \leq 1$, and the sum of the probabilities of the outcomes in the sample space equals 1.

   • The coin is *fair*, which means that each of the outcomes in the sample space has the same probability. In other words, there is nothing about the coin or the process of tossing the coin that favors one outcome over the other.

   On the basis of these assumptions, what is the probability of getting heads in the experiment of tossing a fair coin? _____ Justify your answer.

Pascal's Bonanza (continued)

Name _____

2. If an experiment consists of tossing a coin two times, then you can easily determine the sample space of possible outcomes by making a tree diagram. Examine the tree diagram below. Next to it is a list. All the possible outcomes of the experiment have been translated to this list. Finally, to the right of the list is a probability table for the experiment. All the outcomes from the list have been entered into the table. Dotted arrows connect corresponding outcomes in the tree diagram, the list, and the table.

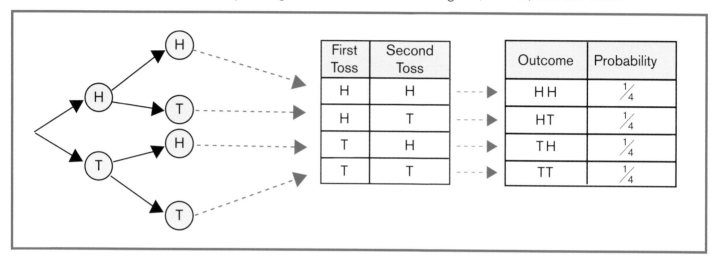

First Toss	Second Toss		Outcome	Probability
H	H		HH	$\frac{1}{4}$
H	T		HT	$\frac{1}{4}$
T	H		TH	$\frac{1}{4}$
T	T		TT	$\frac{1}{4}$

a. Describe how the tree diagram represents the experiment and its possible outcomes.

b. Explain why the probability of each outcome in the sample space is ¼.

c. An *event* is a subset of the outcomes in the sample space of an experiment. For example, in the experiment above, the event "The number of heads is the same as the number of tails" consists of the outcomes HT and TH. The probability of an event is assumed to be the sum of the probabilities of the outcomes that make up the event. In a two-toss experiment, what is the probability that the number of heads equals the number of tails?

Pascal's Bonanza (continued)

Name _____

3. *a.* Suppose that an experiment consists of tossing a coin three times. Create a tree diagram like that in step 2, translate its data into a list of the outcomes in the sample space, and make a probability table for the outcomes. Justify your claims about the probabilities of the possible outcomes.

b. Complete the following table, showing the number of outcomes from the sample space in step 3(*a*) belonging to each of the four events.

Event: Total number of heads	0 heads	1 head	2 heads	3 heads
Number of outcomes				

4. *a.* Compare the number of outcomes in the sample space when the experiment consists of one toss, two tosses, and three tosses of the coin. Use the pattern that you detect to predict the number of outcomes in the sample space when the coin is tossed four times. Justify your prediction.

b. Generalize your results from step 4(*a*) to answer the question, "How many possible outcomes are in the sample space when the coin is tossed *N* times?" _____ Be sure to justify your answer.

5. The list of all possible outcomes of the four-toss experiment is easy to generate if you notice the pattern in the tables for the two-toss and three-toss experiments.

Pascal's Bonanza (continued)

Name _____

a. Describe the pattern and fill in the following table, which lists all possible outcomes of the 4-toss experiment.

1st Toss	2nd Toss	3rd Toss	4th Toss	Probability of Outcome

Explain why all the outcomes are equally likely.

Name _____

b. What is the probability of the event "Exactly half of the four tosses are heads"?

6. *a.* As the number of tosses in Kim's experiment increases, the number of possible outcomes grows exponentially! Kim wants to group the outcomes in the sample spaces by their numbers of heads. In the following table, fill in the row for the 4-toss sample space with the number of outcomes in each event. The rows for the 1-, 2-, and 3-toss experiments already show the outcomes, grouped by numbers of heads, that you generated for these sample spaces in steps 2 and 3.

Event: Total Number of Heads

Experiment	0	1	2	3	4	5	6	7	8	9	10
1-toss	1	1	0	0	0	0	0	0	0	0	0
2-toss	1	2	1	0	0	0	0	0	0	0	0
3-toss	1	3	3	1	0	0	0	0	0	0	0
4-toss											
5-toss											
6-toss											
7-toss											
8-toss											
9-toss											
10-toss											

b. Look in the first four rows of the table for patterns that can help you fill in the row for the 5-toss sample space. Describe a pattern and compare it with the patterns that other students in your class noticed.

c. In the table above, the row for the 5-toss sample space includes a shaded cell for the event "3 heads." Kim notices that to get three heads in the 5-toss experiment, she must get a head on the fifth toss and exactly two heads in the first four tosses *or* get a tail on the fifth toss and exactly 3 heads on the first four tosses. How could she use this observation and the values you have entered for the 4-toss experiment to fill in the shaded cell in the table?

d. Use the pattern that helped you complete step 6(*c*) to complete the rest of the table.

7. *a.* The 10-toss experiment has 2^{10}, or 1024, possible outcomes. In theory, how many of those outcomes would lead to a win for Brenda if she and Kim played the game consisting of 10 tosses of the coin?

b. What is the theoretical probability that Brenda would win the game with Kim if they tossed the coin 10 times?

c. How does this probability compare with the relative frequency of Brenda winning that you found in your simulation of the 10-toss game in Binomial Bonanza?

8. The data that you have in the table in step 6 make up the first rows of what is sometimes called *Pascal's triangle*. These numbers are often arranged in the following manner:

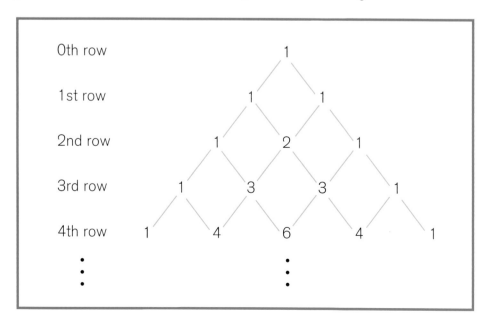

Pascal's Bonanza (continued)

Name _____

As you discovered, you can continue to generate rows in the table (or the triangle) if you have sufficient time — and space — to record the increasingly large numbers that you will get. The Web offers tools that you can use to generate rows efficiently. Go to the following Web site to generate the 20th and 40th rows of the table that you started in step 6:

http://mathforum.org/dr.cgi/pascal.cgi

Use these results to help you repeat step 7 for the 20-toss and 40-toss versions of the game. Report your answers in the boxes below.

20-Toss Game	40-Toss Game
a.	a.
b.	b.
c.	c.

9. *a.* Was Kim justified in not wanting to play the 20-toss game that Brenda originally proposed?

   b. How well do your simulations predict Kim's chances of winning games consisting or 20 and 40 tosses?

Navigating through Reasoning and Proof in Grades 9–12

Derek Jeter Steps into the Box

Name _____

The Rise & Shine Breakfast Company is offering a free 3-by-5-inch baseball card in each box of its popular cereal Sweet & Soggy Squares. Featured on each card is one of the following players: Ken Griffey Jr., Derek Jeter, Pedro Martinez, and Ichiro Suzuki. The company claims that it has randomly distributed equal numbers of the cards in boxes of Sweet & Soggy Squares sold all across the country.

Many children in New York want a Derek Jeter card. On average, how many boxes of the cereal would a family of Derek Jeter fans need to buy to get a Derek Jeter card?

1. Estimate the average number of boxes that a family would need to buy to obtain a Derek Jeter card.

2. Explain why you think your estimate is reasonable.

3. List any assumptions that you made in determining your estimate.

4. *a.* Describe a method that you think would be effective for checking the accuracy of your estimate.

   b. Compare your method with those of other members of your class.

5. Statisticians use simulations to study averages like the one in this problem. You can simulate the baseball-card result of buying a box of Sweet & Soggy Squares by randomly generating a number from 1 to 4 with a calculator, a tetrahedral (four-sided) die, or a spinner with four equal areas. Each number must have an equal chance of occurring and will stand for one of the cards—for example, 1 = Griffey; 2 = Jeter; 3 = Martinez; and 4 = Suzuki. One trial of this simulated experiment will consist of generating a list of random integers from 1 to 4 until you get a 2. The number of integers in your list will represent the number of boxes that a hypothetical family has to buy to get a Derek Jeter card.

Derek Jeter Steps into the Box (continued)

Name _____

a. Conduct 30 trials of your simulation. In each trial, record the list of integers that you obtain on the accompanying data sheet for "Derek Jeter Steps into the Box."

b. After 30 trials, create a frequency table of the results on the second page of the data sheet.

6. In your 30 trials, what was the average length of a trial? _____ In other words, on average, how many boxes of cereal would someone need to purchase to get a Derek Jeter card? _____ How does this compare with your estimate in step 1?

7. Pool your results with those of five of your classmates, so that you have the results of 180 trials in all. Use the frequency chart below to record the pooled results. Complete the chart, extending it if necessary.

Average Number of Boxes Necessary to Obtain a Derek Jeter Card
(30 Simulations)

	1	2	3	4	5	6	7	8	9	10	11	12	13	14	15	16	17	18	19
Student 1																			
Student 2																			
Student 3																			
Student 4																			
Student 5																			
Student 6																			
Column totals																			
Relative frequency																			

8. When averaging the lengths of the 30 trials, you computed the following ratio:

$$\text{Average length of a trial} = \frac{\text{Sum of the lengths of the 30 trials}}{\text{30 trials}}$$

Derek Jeter Steps into the Box (continued)

Name _____

a. Relate the numerator to the total number of boxes purchased in the 30 trials.

b. Relate the denominator to the total number of Derek Jeter cards obtained in the 30 trials.

c. Use these relationships to express a relationship between the relative frequency of obtaining a Derek Jeter card and the average length of a trial.

d. Calculate the average length of a trial in your pooled results of 180 trials and then repeat steps 8(a)–8(c), using the pooled data from 180 trials.

9. a. One-fourth of all boxes of Sweet & Soggy Squares contain a Derek Jeter card. This means that the probability of getting a Derek Jeter card in a random purchase of a box of cereal is ¼. It also means that there is a high probability that the relative frequency of purchased boxes containing Derek Jeter cards will get arbitrarily close to ¼ as the number of purchases increases. Theoretically, what is the average number of boxes that someone needs to purchase to get a Derek Jeter card? _____ Justify your answer.

b. How accurate were the results of your 30-trial simulation in estimating the average number of boxes that someone needs to purchase *in theory* to get a Derek Jeter card? How accurate were your results from the 180-trial simulation? Explain your answers.

Name _____

10. *a.* Use statistical methods to analyze the relative frequencies in your pooled data set from step 7. You can make graphs, calculate mean and standard deviation, and so on. Consider the range of relative frequencies in the pooled data. If someone purchases four boxes of Sweet & Soggy Squares, what is his or her chance of not getting a Derek Jeter card? _____ Or if someone wants to make only one trip to the store, how many boxes should he or she buy to have a 90 percent chance of getting a Derek Jeter card?

b. Use a tree diagram or some other method to find the theoretical probability that a family gets its first Derek Jeter card in the fourth box of cereal purchased. How does this probability compare with the experimental probability of this event that you obtained from the pooled 180-trial simulation?

Data Sheet for
"Derek Jeter Steps into the Box"

Name _____

Number of Trial	List of Integers																					Length of Trial
1																						
2																						
3																						
4																						
5																						
6																						
7																						
8																						
9																						
10																						
11																						
12																						
13																						
14																						
15																						
16																						
17																						
18																						
19																						
20																						
21																						
22																						
23																						
24																						
25																						
26																						
27																						
28																						
29																						
30																						

Data Sheet for "Derek Jeter Steps into the Box" continued

Name _____

Frequency Table of Number of Trials Resulting in the Given Lengths

Length of trial	1	2	3	4	5	6	7	8	9	10	11	12	13	14	15	16	17	18	19
Frequency																			

Late Shift at the Orange Bowl

Name _____

The last bus of the day from the Old Town Line in Miami arrives at the Orange Bowl between 11:00 and 11:20 p.m. The last bus from the New Town Line arrives between 11:10 and 11:25 p.m. The exact arrival times of each bus are randomly and uniformly distributed over its time interval. Tony and Rhonda both work the late shift at the Orange Bowl. Tony always arrives on the last Old Town bus, and Rhonda always arrives on the last New Town bus. Today is Rhonda's birthday, and Tony wants to surprise her with flowers as she climbs off her bus. What is the probability that Tony's bus will arrive before Rhonda's?

1. *a*. Design a simulation of the arrival of the two buses and describe how to use your simulation to make an estimate of the probability that Tony arrives before Rhonda.

   b. What assumptions have you made in designing your simulation?

2. *a*. As a class, discuss the designs of several simulations. Decide on one design that everyone will use independently to estimate the probability that Tony will arrive before Rhonda. On your own, run this simulation to generate 100 trials. Record the number of times that Tony's simulated arrival time falls ahead of Rhonda's simulated arrival time.

   Number of times (in 100 trials) that Tony arrives first _____

   b. On the basis of the results of your simulation, calculate an experimental probability for the event "Tony arrives at the Orange Bowl before Rhonda."

   Experimental probability (based on the results of 100 trials) _____

3. *a*. Each of your classmates also conducted the simulation. Were all the experimental probabilities the same? _____ Why, or why not?

   b. Combine the results of all your classmates' simulations to make an estimate of the theoretical probability. Also make an estimate of the uncertainty of your prediction.

   Estimate of the theoretical probability that Tony arrives first _____

   Estimate of the uncertainty of the probability estimate above _____

4. *a*. To calculate the theoretical probability of Tony arriving before Rhonda, you can think of each possible outcome as an ordered pair (Tony's arrival time, Rhonda's arrival time) and use geometry to represent

Name _____

the set of all possible outcomes. Time is a continuum of moments, so listing all the pairs of arrival times that could occur is impossible. However, using the grid below, draw and label a region representing the set of all the possible pairs of arrival times of the two buses.

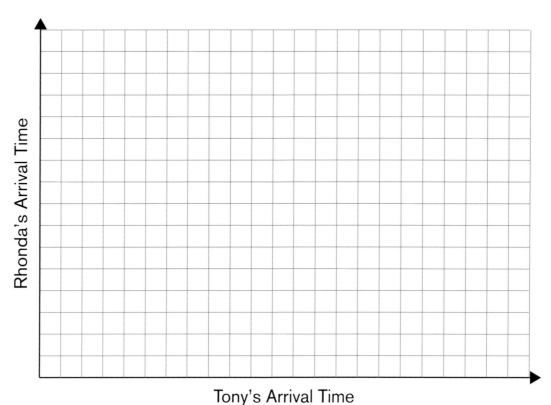

b. What is the area of the region that you have drawn?

Area of the region representing all the possible outcomes _____

5. *a.* In the region that you constructed in step 4, shade the portion that corresponds to the outcomes in which Tony arrives before Rhonda. What is the area of this shaded region?

Area of the shaded region _____

b. Calculate the theoretical probability that Tony arrives before Rhonda, and explain any assumptions that you are making in your calculation.

Theoretical probability that Tony arrives first _____

Name _____

6. Write a paragraph in which you compare or contrast the theoretical result that you obtained in step 5 with the estimate that you made in step 3 on the basis of the pooled results. How accurate were the simulations?

7. Create another probability question involving the arrival times of Tony and Rhonda. Your problem should be one that you can solve by using the geometric method described in step 4. Show the solution to the problem.

Pushing the Perimeter

Name _____

Kelsey's class conducted an investigation of area and perimeter. Afterward, Kelsey made a conjecture:

When the perimeter of a rectangle increases, then the area must also increase.

Kelsey produced the following work in support of her conjecture:

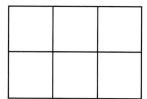

Perimeter = 10 units
Area = 6 square units

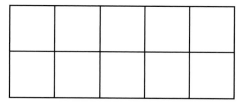

Perimeter = 14 units
Area = 10 square units

1. Do you think Kelsey's conjecture is correct? _____ In one or two sentences, say why or why not?

2. Investigate Kelsey's claim. To organize your investigation, complete the chart below, entering dimensions for rectangles of your own choosing. Use the following questions to guide you in your exploration:

   a. Would considering squares as a particular subset of rectangles help you test Kelsey's conjecture?

   b. How could you organize the data in the chart to help you find patterns or trends?

Dimensions of Rectangle	Shape of Rectangle (Drawing)	Perimeter of Rectangle	Area of Rectangle
2 × 3		10	6

Pushing the Perimeter (continued)

Name _____

3. Answer the following questions on the basis of your data:

   a. What patterns do you notice in your data?

   b. Did you investigate any rectangles whose perimeters are the same but whose areas are different? _____ If so, what are the perimeters and areas of these rectangles?

   c. Did you investigate any rectangles whose areas are the same but whose perimeters are different? _____ If so, what are the areas and perimeters of these rectangles?

   d. If the area of a rectangle is 100 square units, what is the largest perimeter that the rectangle can have? _____ What is the smallest perimeter it can have?

4. Reconsider Kelsey's conjecture:

When the perimeter of a rectangle increases, then the area must also increase.

   What is your evaluation of the claim now? Is it true or false?

5. Can you justify your response in step 4? If you think the conjecture is true, try to find a convincing way to prove its truth to your classmates. If you think the conjecture is false, look for a convincing way to demonstrate that conclusion. *Hint:* To make your case—

   • organize your evidence in a way that you think will make sense to others;

   • present your argument in a clear, easy-to-understand fashion;

   • make your argument as general as possible so that it will persuade others that your answer is *always* correct.

Name _____

6. *a.* Would Kelsey's conjecture hold for other shapes, such as regular and irregular polygons in general?

b. Would Kelsey's conjecture hold for circles?

c. How are the situations for polygons in general and circles the same as or different from the situation for rectangles that you explored above?

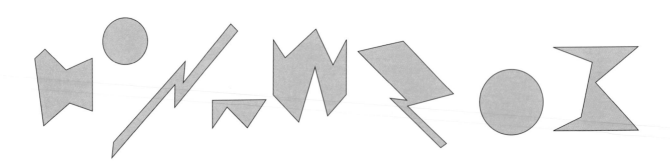

Navigating through Reasoning and Proof in Grades 9–12

Heavy Metal at Maximum Volume

Name _____

Maximizing the Measurement – Part 1

Suppose that the members of your group are employees of Blue Metal Designs. Management is planning to make a new open-topped box and market it under the name "Heavy Metal." This box will be manufactured from custom-cut sheet metal, 16 cm × 30 cm, and will have the largest possible volume that these dimensions allow. Management has selected your team to design the Heavy Metal box and has given you a piece of the sheet metal and the diagrams below, showing how the company creates an open-topped box by cutting out four equal-sized squares from the four corners of the sheet metal.

After cutting out the squares shown on the left, metalworkers use machines to fold the metal on the dashed lines shown on the right so that the four smaller rectangles outside the lines form the sides of the box.

1. *a.* As a team, guess how large the cut-out squares must be to make the Heavy Metal box with maximum volume from sheet metal precut to 16 cm by 30 cm.

   b. Work with construction paper or tagboard to construct a full-size model of your team's box.

   c. Measure the volume of your box and compare it with that of boxes made by other teams.

2. As a team, refine your guess on the basis of your comparisons in 1(*c*). Make a new prediction of the size of the square that Blue Metal Designs should remove from each corner of the sheet metal to make the new Heavy Metal box. Record your prediction, and explain your reasoning.

   Our prediction: _____

Heavy Metal at Maximum Volume (continued)

Name _____

3. One way to explore the box of maximum volume is to draw diagrams and use them to test the range of possibilities. You can record successive observations in a table like the following.

Side Length of Cut-Out Square (cm)	Length of Resulting Box (cm)	Width of Resulting Box (cm)	Height of Resulting Box (cm)	Volume of Resulting Box (cm^3)
1	14	28	1	392
2	12	26	2	624
3	10	24	3	720
⋮	⋮	⋮	⋮	⋮
x				

Work together with your team to make such a table with spreadsheet software or by using the **function editor** and **table** features of a graphing calculator. Answer the questions below to help your group set up and complete a table:

a. The values in column 1 indicate the side lengths of squares to be cut from the corners of the sheet metal. What is the range of possible values for the side lengths? _____ Why?

b. The values in column 1 determine all the values calculated in the other columns. By what increment will your group change the side length of the cut-out square in your investigation? _____ By 1 cm each time? By 0.5 cm? By 0.1 cm? How can you choose, and why?

c. How can you use the values in column 1 to determine the values in columns 2 through 4 of the table?

Name _____

d. How can you calculate the value to enter in column 5 for the volume of the resulting box?

4. Your team can use different methods to "narrow in" to refine your estimate of the maximum volume of the box. Narrow in first by working with a graph of the volume function:

   a. Think about the sequence of calculations that you make to fill in a row of the table, beginning with a side length for the cut-out square in column 1 and ending with the volume of the resulting box in column 5.

   • For a side length *x*, how would you fill in the rest of the columns for the row?

   • In column 5, what value would you enter for the volume in terms of *x*:

   $$V(x) = \underline{\hspace{3cm}}.$$

   • This equation reflects the sequence of operations that you use to obtain the entries in each row in the table.

   b. On your calculator or spreadsheet, create a graph of the volume function that you wrote above.

   c. How can you use the graph to estimate the maximum possible volume of the box?

   d. On the basis of your team's graph, what is your best estimate of the maximum volume of the box? _____ Could there be more than one maximum volume? _____ Why, or why not?

5. Now "narrow in" on an estimate of the maximum volume in a different way. This time, use successively smaller ranges and increments in a series of electronic tables.

   a. Create a series of at least three tables that use successively smaller ranges and smaller increments within each range for the *x*-values in column 1. For example, you might start with *x*-values between 0 and 8 with increments of 1, and then move to a range between 2 and 5 with increments of 0.3. What ranges and increments will you use, and why?

   b. What maximum value do you find in each of your three successive tables?

Name _____

c. On the basis of your team's tables, what is your best estimate of the maximum possible volume of the box?

6. a. Compare the results from the two methods of narrowing in on the maximum volume. On the basis of your work with both methods, what is your best estimate of the maximum possible volume of the open-topped Heavy Metal box?

b. Which of the two methods of "narrowing in" do you think provides a better estimate of the maximum volume? _____ Explain.

7. Use your group's work in steps 2–6 to write a convincing justification that you have found the best estimate of the maximum volume of the Heavy Metal box. Your justification should also discuss the amount of uncertainty (or possible error) in your estimate.

Not Only Useful, but Also Orna-Metal

Name _____

Maximizing the Measurement – Part 2

Blue Metal Designs also sells boxes with tops. The company customarily puts ornamental trim around the edge of the tops of such boxes. Blue Metal Designs imports the metal trim from Belgium, in very narrow one-meter lengths.

Management has selected your team for a new project. This time you must determine the dimensions of the box top of maximum area that Blue Metal Designs can edge with exactly one piece of decorative trim. In other words, if metalworkers fold the trim to form a rectangle, what is the box top of largest area that they can edge with it? The drawing below shows one way of folding the trim, with the folds occurring at the 10-centimeter, 50-centimeter, and 60-centimeter points on a trim piece aligned with a meter stick.

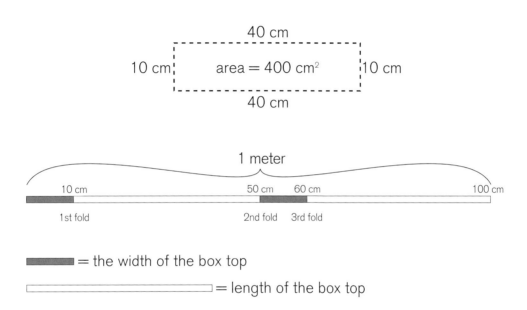

1. Make a prediction about the dimensions of the box top of largest area that Blue Metal Designs can edge with one piece of trim. Record your prediction and explain your reasoning.

Not Only Useful, but Also Orna-Metal (continued)

Name _____

2. Working as a group, use a computer with spreadsheet software or the function editor and table features on a graphing calculator, and set up a table like the one below.

Width of Box Top (cm) (Length of Trim to First Fold)	Length of Box Top (cm) (Length of Trim between First and Second Folds)	Area of Rectangle (cm^2)
1	49	49
2	48	96
3	47	141
⋮	⋮	⋮
x		

Such a table can help you explore the range of possible box tops. Answer the questions below to help your group set up and complete a table.

a. Consider the first two columns of the table. What is the range of possible values for the length of trim to the first fold? _____ How is this length related to the length of trim between the first and second folds?

b. At what point along the trim does the second fold always occur? _____ Why?

c. How can you use the values in column 1 in the table to compute the values in columns 2 and 3?

3. Work with your group to "narrow in" on the maximum area by using successively smaller ranges and smaller increments in the first column of your table.

Name _____

a. From your data, what is your best estimate of the maximum rectangular area that Blue Metal Designs can edge by one piece of trim?

b. What is the uncertainty (or largest possible error) in your estimate?

4. Let the length of the trim to the first fold — the value in column 1 of your table — be *x*.

a. Fill in columns 2 and 3 of your table in terms of *x*. In column 3, you should get a value for the area in terms of *x*. Use this value to write an equation for the rectangular area of the box top as a function of *x*.

$$A(x) = \underline{\hspace{5cm}}$$

b. Graph this area function and zoom in on an estimate of the maximum rectangular area that Blue Metal Designs can edge with exactly one piece of trim. On the basis of your graph, what is your best estimate for the maximum area?

c. What is the uncertainty (or largest possible error) in your estimate?

5. *a.* Compare your results from the two methods of narrowing in to make your best estimate of the box top of maximum area that Blue Metal Designs can edge with exactly one piece of trim.

b. Use the function that you generated in step 4(*a*) to prove that you have indeed found the maximum value. *Hint:* What is the shape of the graph of the function, and what do you know about finding the maximum or minimum point on such a graph?

c. In general, what can you say about the rectangle of maximum area that can be enclosed by a perimeter of a particular length?

6. Blue Metal Designs uses the same ornamental trim to edge the company's custom-made rectangular trays. If a client orders a rectangular tray and specifies only that the area be 900 cm^2, then what is the minimum length of trim that Blue Metal Designs will need to edge the tray? Work with the members of your team to solve this problem. Then present your group's solution on a poster, showing any tables, graphs, or other evidence that might convince the managers of Blue Metal Designs that your solution uses the smallest amount of trim to enclose an area of 900 cm^2.

Solutions for the Blackline Masters

Solutions for "Probing Products"

1. *a.* Answers will vary, but most students should notice that the products are all even numbers.

    b. A sample "if … then" statement follows: "If two consecutive whole numbers are multiplied, then their product is an even number."

    c. The consecutive numbers *j* and *k* cannot both be even since the difference between two distinct even numbers is at least 2, and the difference between *j* and *k* is only 1. With appropriate modifications, the same argument can be used to prove that *j* and *k* cannot both be odd.

    d. One number must be even, and the other must be odd.

    e. The product of two whole numbers is even when either or both of the whole numbers is even. Since either *j* or *k* must be even, then their product is even.

2. In an array of $n(n+1)$ dots, the dots can be divided into two congruent right triangles of dots:

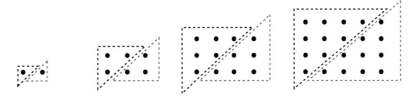

    This demonstrates that the product $n \times (n+1)$ is two times some number (the number of dots in one of the triangles) and is hence an even number. Another visual argument could be based on the fact that in an array representing the product $n(n+1)$, either the number of rows or the number of columns is even. Thus, the array can be partitioned into two parts with the same number of rows or the same number of columns:

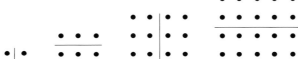

3. *a.* Answers will vary. "Squares of even numbers are even numbers" and "Squares of even numbers are multiples of 4" are likely responses.

    b. Arrays should show the product of $2k$ and $2k$. For example, the product could look like the product of $(k+k)$ and $(k+k)$, which is a square made up of four smaller k^2 squares. The illustration below shows a 6×6 dot array divided into four 3×3 dot arrays:

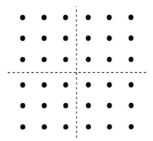

c. Let *n* be an even number. Therefore, $n = 2k$ for some integer *k*. Thus, $n^2 = (2k)^2$, which simplifies to $4k^2$, a multiple of 4.

4. *a.* Answers will vary. "Squares of odd numbers are odd numbers," "Squares of odd numbers are one more than a multiple of four," and "Squares of odd numbers are one more than a multiple of eight" are all likely responses.

b. Methods of constructing visual arguments will vary. One approach is to draw four congruent rectangles on a dot array that shows $(2k + 1)^2$ and enclose $k \times (k + 1)$ dots in each rectangle, excluding the dot in the center of the array from every rectangle, as shown on the right. Each rectangle can then be cut by a diagonal, showing a total of eight congruent triangles with one dot left over in the center. The representation thus shows that one less than the square of an odd number is divisible by 8.

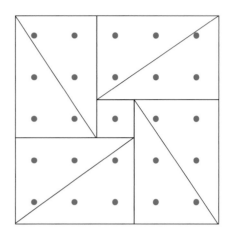

Another method uses a diagram that shows the four rectangles without the diagonal cuts. Students will note that the odd number $2k + 1$ can be rewritten as $k + (k + 1)$, whose square is $(2k + 1)^2 = [(k) + (k + 1)]^2$, and can be represented as shown on the right.

Each of the rectangular subregions has sides of *k* and $(k + 1)$. The students discovered earlier that the product $k \times (k + 1)$ is equal to twice a triangular number T_k. Hence,

$$(2k + 1)^2 = 4k^2 + 4k + 1$$
$$= 4k(k + 1) + 1$$
$$= 4(2T_k) + 1$$
$$= 8T_k + 1.$$

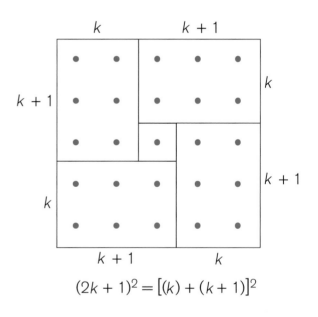

$$(2k + 1)^2 = [(k) + (k + 1)]^2$$

c. Let *n* be an odd number. Therefore, $n = 2k + 1$, where *k* is an integer. Then one less than the square of *n* is $(2k + 1)^2 - 1$, or $4k^2 + 4k + 1 - 1$, or $4k^2 + 4k$. By factoring, we get $4k(k + 1)$, or $4[k(k +1)]$. This number is clearly divisible by 4, but it is also divisible by 8 because $k(k + 1)$ is the product of consecutive numbers, which we showed earlier to be even.

5. Answers will vary. The representations above for 4(*b*) show that when *n* is odd, $n = 2k + 1$, and the *n*th square number (n^2) is one more than 8 times the *k*th triangular number.

Solutions for "Regions of a Circle"

1–3. For 4 points on a circle, the maximum number of regions created is 8. For 5 points on a circle, the maximum number of regions created is 16 (see the illustrations in the margin on p. 14).

4. Answers will vary. Students are likely to observe the doubling pattern. They may express this pattern explicitly or recursively. For example, they may explicitly state, "If a circle has n points on it, then the maximum number of regions formed is 2^{n-1}. Or they may offer a recursive statement: "If a circle has n points on it, then the maximum number of regions formed, R_n, is $2(R_{n-1})$, or two times the number of regions formed by a circle with $(n-1)$ points identified on it."

5. Students may appeal to the pattern observed in the first five cases; however, this pattern does not hold beyond 5 points.

6. *a* and *b*. Students should discover that 6 points on a circle create a maximum of 31 regions.

   c. To maximize the number of regions, the students must draw chords so that no three chords intersect at a common point in the interior of the circle.

   d. Showing that identifying 6 points on a circle and drawing all possible chords creates a maximum of 31 regions generates a counterexample to the conjecture about the doubling pattern. For 7 points, the maximum number of regions is 57. For 8 points, the maximum number of regions is 99. It may be difficult for students to see a pattern here. However, if they do, encourage them to share their conjectures and challenge others to justify or generate counterexamples to refute these conjectures. The maximum number of regions that can be formed by chords connecting n points on a circle is $\binom{n}{4}+\binom{n}{2}+1$.

   Students who conduct an Internet search on the key words *geometry*, *partitioning*, and *circle* can find an explanation of why this formula works at http://www.cut-the-knot.org (on the Geometry page under "Partitioning a Circle"), or you can send them directly to http://www.cut-the-knot.org/Generalization/cuttingcircle.shtml.

7 *a.* Most students will suggest a pattern such as $A_n = 3n$. Some might conjecture that A_n equals $n^2 + 2$. These patterns generate different tables, as shown below.

n	1	2	3	4	5
$A_n = 3n$	3	6	9	12	15

n	1	2	3	4	5
$A_n = n^2 + 2$	3	6	11	18	27

   b. Suppose that students conjectured that A_n equals $3n$. Adding the product $(n-1)(n-2)$ to $3n$ gives the formula $A_n = (n-1)(n-2) + 3n$, which yields the following values for the table:

n	1	2	3	4	5
$A_n = (n-1)(n-2) + 3n$	3	6	11	18	27

   (This formula is the same as $A_n = n^2 + 2$.) If, however, the students originally conjectured that A_n equals $n^2 + 2$, then adding the product $(n-1)(n-2)$ gives the formula $A_n = (n-1)(n-2) + n^2 + 2$, which yields the following values for the table:

n	1	2	3	4	5
$A_n = (n-1)(n-2) + n^2 + 2$	3	6	13	24	39

  c. Dania's formula gives $A_1 = 3$ and $A_2 = 6$. Therefore, it is a valid answer to step 7(*a*).

8. When the values of only A_1 and A_2 are known, infinitely many formulas exist that yield these first two terms as well as infinitely many candidates for A_3. Observe that the formula $A_n = k(n-1)(n-2) + 3n$ yields $A_1 = 3$ and $A_2 = 6$, no matter what the value of k is. Furthermore, $A_3 = 9 + 2k$, and this expression can have infinitely many values, depending on one's choice of k. Thus, if all we know about the sequence are the values of A_1 and A_2, we can say nothing for certain about the value of A_3. Students should discuss their answers to this question. It makes sense to assert that if we know nothing about the process that generated the numbers A_1 and A_2, why couldn't any number that we choose be the value for A_3?

9. *a.* Answers will vary. A student who substitutes -2 for k in the formula, obtaining

$$P_n = -2(n-1)(n-2)(n-3)(n-4)(n-5) + 2^{n-1},$$

  will enter the following values in the table:

n	1	2	3	4	5	6
P_n	1	2	4	8	16	-208

  b. The tables are the same for the first five terms. Once again, simply knowing five numbers doesn't tell you what a sixth number should be. Indeed, if one can find a formula that fits the first five terms of a sequence such as $R_n = 2^{n-1}$, then one can find infinitely many formulas that fit those five terms but differ in the sixth term.

10. Answers will vary. The point to make is something of a truism. You can't make a claim that applies to all the terms of an infinite sequence unless you know something about all the terms of the infinite sequence or the process that generates all the terms. Knowing only the first N terms tells you nothing about the rest of the terms in a sequence.

Solutions for "Algorithmic Aptitude"

1. *a.* The steps in the teener multiplication algorithm are shown below:

$$13 \times 19 \rightarrow 13 + 9 = 22 \rightarrow 10 \times 22 = 220 \rightarrow 220 + 3 \times 9 = 220 + 27 = 247$$
$$16 \times 18 \rightarrow 16 + 8 = 24 \rightarrow 10 \times 24 = 240 \rightarrow 240 + 6 \times 8 = 240 + 48 = 288$$
$$17 \times 12 \rightarrow 17 + 2 = 19 \rightarrow 10 \times 19 = 190 \rightarrow 190 + 7 \times 2 = 190 + 14 = 204$$

  The students make up two problems of their own to test the algorithm, and they should discover that the algorithm gives the correct answer in every case.

  b. Using the teener multiplication algorithm to multiply two teener numbers expressed as

$$10 + k \text{ and } 10 + h,$$

where k and h are both one-digit numbers from 1 through 9, gives the following result:

$$(10 + k) \times (10 + h) \rightarrow (10 + k) + h = 10 + k + h \rightarrow 10 \times (10 + k + h) =$$

$$100 + 10k + 10h \rightarrow (100 + 10k + 10h) + (k \times h) = 100 + 10k + 10h + kh$$

$c.$ Using traditional algebraic techniques to multiply $10 + k$ and $10 + h$ yields the following:

$$(10 + k)(10 + h) = 100 + 10k + 10h + kh$$

This result agrees with that in 1(b). Students should discuss their work in 1(b) and 1(c) to make sure that they understand that the agreement of the teener result with the conventional result validates the teener algorithm for every case of k and h.

2. $a.$ Ted's second and fourth answers are incorrect. Ted apparently subtracts the smaller number from the larger in a column, regardless of the numbers' vertical positions. Since Ted is subtracting a smaller two-digit number from a larger one in every case, the flaw in his method produces an error when the ones digit of the subtrahend is larger than the ones digit of the minuend. The result of each flawed subtraction is always greater than the actual difference between the two numbers. When Ted uses his method in balancing his checkbook, his results will make him think he has more money left in his account than he really does, and soon he will be bouncing checks!

$b.$ Students' conjectures will vary about the largest error that Ted can make in subtracting a smaller two-digit amount from a checking account balance from \$10 to \$99. Examples will be similar to the following:

$$
\begin{array}{r}
87 \\
- 39 \\
\hline
52
\end{array}
\text{ (Correct answer 48; error of +4)}
\qquad
\begin{array}{r}
91 \\
- 49 \\
\hline
58
\end{array}
\text{ (Correct answer 42; error of +16)}
$$

However, note that if Ted's balance is a multiple of 10, the difference can be maximized. In fact, the largest error that Ted can make is an error of \$18. An example follows:

$$
\begin{array}{r}
80 \\
- 39 \\
\hline
59
\end{array}
\text{ (Correct answer 41; error of +18)}
$$

The maximum possible error is +18, and it occurs when the minuend has a 0 in the units place and the subtrahend has a 9 in the units place.

$c.$ Ted's two-digit checking account balance is $n \times 10 + r$, and $t \times 10 + s$ is a smaller two-digit amount that Ted is subtracting from it; n and t are whole numbers from 1 to 9, and r and s are whole numbers from 0 to 9. An error will result in Ted's subtraction whenever $r < s$ in the ones column. Note that we can assume $n > t$ in all cases, since the amount that Ted is subtracting is less than the balance in his account.

$d.$ Represent the two numbers as $n \times 10 + r$ and $t \times 10 + s$ and subtract:

$$
\begin{array}{r}
n \times 10 + r \\
- (t \times 10 + s) \\
\hline
\end{array}
$$

If $r > s$, Ted's method of subtraction will yield the correct expression:

$$(n - t) \times 10 + (r - s).$$

If $r < s$, Ted's result will be $(n - t) \times 10 + (s - r)$, which is incorrect. This incorrect answer is greater than the correct answer, since $(s - r) > (r - s)$ when $r < s$. In fact, the difference between the incorrect answer and the correct answer is the following:

$$[(n - t) \times 10 + (s - r)] - [(n - t) \times 10 + (r - s)] =$$
$$(s - r) - (r - s) =$$
$$2(s - r).$$

e. The expression $2(s - r)$ has the maximum value when the difference between s and r is as great as possible. Since s and r are one-digit numbers, this will occur when $s = 9$ and $r = 0$. The maximum difference is therefore $2(9 - 0)$, or 18.

Some students might point to an example such as $42 - 17$ and note that when doing the subtraction correctly, one column at a time, they would get $10(4 - 1) + (2 - 7)$, or $30 + -5$, instead of the incorrect answer of $30 + 5$, which Ted's method would yield. The students might observe that the error in Ted's result in this example is the difference between -5 and 5. They might also remark that the error introduced by Ted's method *in general* is the difference between a single-digit number and its opposite. So the worst error that Ted can make is the difference between -9 and 9, or 18. Thus, the minuend would have to end in a zero and the subtrahend would have to end in a 9 for Ted's error to be maximized.

3. All of Marie's answers are incorrect. Marie appears to be merging her multiplication and addition algorithms. She correctly adds the digits in the ones column and renames a ten as needed. She then adds the single-digit addend to the digit in the tens column in the addend above. Finally, she adds on the renamed ten to the tens digit:

$$\left(\begin{array}{c} 10a + b \\ + \quad c \end{array} \right) \longrightarrow \boxed{(b + c) + 10(a + c)} \longrightarrow \boxed{10a + b + 11c}$$

The correct algorithm yields the expression $10a + b + c$. Marie's result of $10a + b + 11c$ is off by the amount $10c$. Thus, the maximum error occurs when $c = 9$, for a maximum error of 90.

4. A variety of situations and algorithms are possible. In the case of dividing a two-digit number by 5, students may invent a procedure such as "double the number and move the decimal point one place to the left." Verification of the general case proceeds as follows:

$$N \div 5 = \frac{N}{5} = \frac{2 \times N}{2 \times 5} = 2N \div 10.$$ To divide by 10, one moves the decimal point one place to the left.

Solutions for "Demystifying Magic Squares"

1. The magic sum for the 4×4 square is 34.

2.

8	1	6
3	5	7
4	9	2

The magic sum for this square is 15.

a. The 3 cannot be paired in any row, column, or diagonal with 1 or 2 (note that 3 + 1 or 3 + 2 leaves no third number to create a sum of 15). Therefore, it must be placed in the second row, first column. This also explains why the 3 cannot be placed in the center.

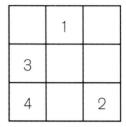

b. The 4 cannot be in any row, column, or diagonal with 1 (note that 4 + 1 leaves no third number to create a sum of 15). This removes the option of the center. Since the 2 and 3 are already placed, two choices remain. Placing the 4 in the second row, third column, would force 8 into the center square and 9 into the first row, third column, creating a di-agonal whose sum is greater than 15. This would violate the fact that the sums of diagonals must be 15. Therefore, we have only one choice for placing the 4. It must go in the third row, first column.

c. Assume that the 1, 2, 3, and 4 are now fixed. To demonstrate that none of the numbers 6, 7, 8, or 9 can occupy the center square, consider the following. Placing 9 in the center requires a 3 (already used) to complete the second row. The 8 must go in the upper left to complete the first column. Placing 7 in the center is impossible, since another 7 would be needed to complete the second column. Similarly, placing 6 in the center would require another 6 to complete the second row.

d. A process of elimination identifies 5 as the number that must occupy the center square.

3. *a–b.* The students should discover that they cannot make the first and second squares "magic." The 1 cannot occupy any corner or the center position, since such a placement would require that it be part of at least three sums adding up to 15. This is not possible with the restriction that the other digits be 2 through 9, without repetition. "Magic" solutions for the remaining squares are shown below:

Not possible | Not possible

	1	

1		

6	7	2
1	5	9
8	3	4

4	9	2
3	5	7
8	1	6

6	1	8
7	5	3
2	9	4

4	3	8
9	5	1
2	7	6

b. A variety of responses are possible. Students' observations may include the following:

- The successful 3 × 3 magic squares always have a 5 in the center position.
- The 1 cannot occupy the center or any corner position in a successful square.
- Successful solutions are reflections or rotations of one another (there is only one successful arrangement of numbers, but it appears in various transformations).

4. *a.* The magic square on the left below contains the entries 5, 6, 7, 8, …, 13, and the square on the right contains the entries 2, 4, 6, 8, …, 18.

12	5	10
7	9	11
8	13	6

Magic sum
= 27

16	2	12
6	10	14
8	18	4

Magic sum
= 30

Students' selections of entries for the third square shown on the activity sheet will vary.

b. The sum of the nine entries in a 3 × 3 magic square is three times the magic sum.

c. If the magic sum is S, then the sum of each row, column, or diagonal is S. Using the entries for the rows given in the generalized square on page 105, we have $(a + b + c) + (d + e + f) + (g + h + i)$ $= S + S + S = 3S$. Therefore, the magic sum is 1/3 of the sum of all nine entries in the magic square.

5. *a.* The average of the nine entries in any 3 × 3 magic square is equal to the entry in the center.

b. The students need to prove that the average of the nine entries in a 3 × 3 magic square is equal to the entry in the center of the square, represented by e. The hint on the activity sheet suggests that the students combine the equations $S = d + e + f$, $S = a + e + i$, $S = b + e + h$, and $S = c + e + g$ to show that e is one-third the magic sum S. Applying the hint gives the following:

$$(d + e + f) + (a + e + i) + (b + e + h) + (c + e + g) = 4S.$$

Regrouping the variables yields

$$(a + b + c + d + e + f + g + h + i) + 3e = 4S.$$

The students can simplify this equation by naming the sum $(a + b + c + d + e + f + g + h + i)$. Suppose they call it T, for example. Substituting T in the equation above gives $T + 3e = 4S$. But from step 4(*c*) above, the students know that $T = 3S$. By using this result and substituting, the students will obtain $3S + 3e = 4S$, or $3e = S$, or $e = S \div 3$, which is the average of the nine numbers in the magic square.

6 *a.* Given the set of numbers {1, 3, 5, 7, 8, 10, 12, 14, 16}, the students are likely to try to form a magic square. However, when they try to determine the magic sum by adding all the numbers and dividing by 3, they will quickly realize that the magic sum would have to be 76 ÷ 3, which is not an integer. This quotient cannot be the sum of any three of the integers in the list, so a magic square is not possible with these nine numbers.

b. To explain why the set of numbers {0, 2, 3, 4, 5, 7, 8, 9, 10} cannot be arranged to form a magic square, students are likely to begin by finding a magic sum, which in this case would seem to be 24. The set of numbers offers various choices of three numbers whose sum is 24. Thus, the situation is unlike that in

6(a). However, the average of the nine entries is $\frac{48}{9}$, which, by what the students demonstrated in 5(a), must be the entry in the center square. Since $\frac{48}{9}$ is not one of the entries, the students cannot form a magic square. Alternatively, some students might argue that a 3 × 3 magic square must have eight combinations of three numbers whose sum is 24, but with the given entries, they could come up with only two combinations: $5 + 9 + 10$ and $7 + 8 + 9$.

c. Sets of numbers and explanations will vary.

7. Andy claims that because the center square in a 3 × 3 magic square is always one-third of the magic sum, the four other numbers on the diagonals must be paired so that the center number is the average of each pair. Andy's claim is correct. The students can demonstrate this fact by letting the magic sum S be the sum of the three entries on a diagonal where the center entry is e. The students know from their work in step 5(b) that $e = \frac{1}{3}S$. So the other two entries must sum to $\frac{2}{3}S$ or $2e$. This can happen only if they are of the form $(e + x)$ and $(e - x)$. The average of these two entries is

$$\frac{(e+x)+(e-x)}{2} = \frac{2e}{2} = e.$$

8. The values that the students should enter in the first magic square are shown in boldface below:

$a + x$	**$a - x - y$**	$a + y$
$a - x + y$	a	**$a + x - y$**
$a - y$	**$a + x + y$**	$a - x$

The students should complete the second square as follows:

11	3	10
7	8	9
6	13	5

9. No, Andy's method does not always work. The method depends on the choice of x and y. If $x = y$, $x = 0$, or $y = 0$, then various entries in the square are equal. Such a situation violates the definition of a magic square.

10. Yes, all 3 × 3 magic squares are generated in the fashion described by Andy. The argument in the solution above for step 7 applies to the entries in any row, column, or diagonal of a 3 × 3 magic square. The converse of this result is not true. If a square has the structure identified by Andy, then it does not have to be a magic square. See the solution for step 9.

Solutions for "Grappling with Growth Rates"

1. *a.* The completed tables for the fees at the High Rentz and Rising Sums hotels follow:

High Rentz Hotel	
Length of Stay	Total Cost
1 day	$6,000
2 days	$7,000
3 days	$8,000
4 days	$9,000
5 days	$10,000
6 days	$11,000
7 days	$12,000

Rising Sums Hotel	
Length of Stay	Total Cost
1 day	$300
2 days	$600
3 days	$1,200
4 days	$2,400
5 days	$4,800
6 days	$9,600
7 days	$19,200

b. A function that gives the total cost of renting the rooms at the High Rentz Hotel for *n* days is $H(n) = 5000 + 1000n$.

c. A function that gives the total cost of renting the rooms at the Rising Sums Hotel for *n* days is $R(n) = 300(2^{n-1})$

d. Using the function in 1(*b*), the students should discover that an extended stay of 14 days will cost Sal Snake $19,000 at the High Rentz Hotel:

$$H(14) = 5,000 + 1,000(14) = 19,000.$$

At the Rising Sums, a stay of 14 days will cost Sal $4,915,200:

$$R(14) = 300(2^{13}) = 4,915,200.$$

e. The enormous difference between the cost of a 14-day stay at the High Rentz and a stay of the same length at the Rising Sums reflects the fact that the High Rentz adds a fixed amount to the bill each day, whereas the Rising Sums doubles the charge for each additional day that a person stays. The billing practice at the High Rentz creates a linear relationship between the length of the stay and the cost, but the practice at the Rising Sums creates an exponential relationship between the two variables.

2 *a.* Completed tables for the functions $g(x) = 2x + 1$ and $h(x) = 2^x$ follow:

x	$g(x) = 2x + 1$
1	3
2	5
3	7
4	9
5	11
6	13

x	$h(x) = 2^x$
1	2
2	4
3	8
4	16
5	32
6	64

b. Completed tables for the growth rate functions of $g(x)$ and $h(x)$ follow:

x	$\Delta g(x)$
1	2
2	2
3	2
4	2
5	2

x	$\Delta h(x)$
1	2
2	4
3	8
4	16
5	32

c. The students should note that $\Delta g(x)$ always seems to be 2, and $\Delta h(x)$ always seems to double for consecutive integer values of n. Thus, it appears that formulas for the growth rate functions are $\Delta g(n) = 2$ and $\Delta h(x) = 2^n$, for every natural number n. To prove that these formulas hold in the general case, the students should look at $g(n + 1) - g(n)$ and $h(n + 1) - h(n)$. By substitution,

$$g(n + 1) - g(n) = [2(n + 1) + 1] - [2n + 1] = 2,$$

and

$$h(n + 1) - h(n) = 2^{n+1} - 2^n = (2^1)(2^n) - 2^n = (2 - 1)\,2^n = 2^n.$$

3. *a.* For $k(x) = 4x - 2$, $\Delta k(x) = 4$.

   b. For $r(x) = -3x + 2$, $\Delta r(x) = -3$.

4. *a.* In step 3, $\Delta k(x)$ and $\Delta r(x)$ represent the slopes of the linear graphs of $k(x)$ and $r(x)$, respectively.

   b. Students can state and prove the following generalization: Given a linear function $f(x) = mx + b$, the growth rate function will be $\Delta f(x) = m$. They can prove this generalization in this way: By definition, $\Delta f(x) = f(x + 1) - f(x)$. Substituting gives

$$\Delta f(x) = (m(x + 1) + b) - (mx + b) = mx + m + b - mx - b = m.$$

5. *a.* The growth rate function, $\Delta d(x)$, of $d(x) = x^2$ is $2x + 1$. By definition, the growth rate function is $\Delta d(x) = d(x + 1) - d(x)$. Substituting gives $\Delta d(x) = (x + 1)^2 - x^2 = (x^2 + 2x + 1) - x^2 = 2x + 1$.

   b. Observe the results as the students graph $d(x)$ and $\Delta d(x)$ on their calculators. A screen image of the graphs appears below. The parabola is the graph of $d(x) = x^2$, and the line is the graph of $\Delta d(x) = 2x + 1$.

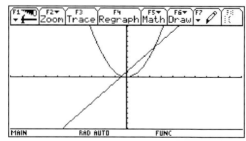

   c. When $\Delta d(x)$ is negative, the graph of $d(x)$ is decreasing from left to right. When $\Delta d(x)$ is positive, the graph of $d(x)$ is increasing from left to right. The value of $\Delta d(x)$ is equal to 0 near the vertex.

6. *a–f.* Observe the results as students enter and plot the required functions. In (*c*), students should note that near, though not precisely at, the vertex of each parabola, the graph of the corresponding linear growth function crosses the x-axis — that is, the rate of growth of Y_1 is 0. In (*d*), the students' entries in the table of values for $\Delta f(x)$ should change sign near the value representing the vertex of the

parabola. In (f), the students should determine that the growth rate function $\Delta f(x)$ for a quadratic function $f(x)$ will always be linear. They can show this by evaluating $\Delta f(x)$ for the general quadratic function $f(x) = ax^2 + bx + c$. By definition, $\Delta f(x) = f(x + 1) - f(x)$. Substituting gives $\Delta f(x) = (a(x + 1)^2 + b(x + 1) + c) - (ax^2 + bx + c)$. Expanding the expressions yields many terms that cancel, leaving $\Delta f(x) = 2ax + a + b$, or $(2a)x + (a + b)$, an expression in the form of a linear function.

7 a. For $t(x) = 2(3^x)$, $\Delta t(x) = 4(3^x)$, and $\dfrac{\Delta t(x)}{t(x)} = 2$.

b. For $q(x) = -2(5^x)$, $\Delta q(x) = -8(5^x)$, and $\dfrac{\Delta q(x)}{q(x)} = 4$.

8. The students should observe that for an exponential function $f(x) = a(b^x)$, the growth rate function $\Delta f(x)$ is also exponential. In addition, they should note that the ratio $\dfrac{\Delta f(x)}{f(x)}$ is a constant value and is equal to $b - 1$. To prove that this observation is true, they can begin with the definition of a growth rate function $\Delta f(x)$ for $f(x)$: $\Delta f(x) = f(x + 1) - f(x)$. By substituting, they obtain $\Delta f(x) = a(b^{x+1}) - a(b^x) = a(b^x)(b) - a(b^x)$. They can simplify this expression by expanding and applying the distributive property:

$$\Delta f(x) = a(b^x)(b) - a(b^x)$$
$$= a(b^x)(b - 1)$$
$$= a(b - 1)(b^x)$$

The students should recognize this last expression as an exponential function with coefficient $a(b - 1)$. The ratio to the growth rate function $\Delta f(x)$ to the function $f(x)$ is thus

$$\frac{\Delta f(x)}{f(x)} = \frac{a(b - 1)(b^x)}{ab^x}$$

The students can cancel common factors to arrive at a constant ratio of $(b - 1)$.

9. To determine whether $p(x)$ is a linear function, a quadratic function, an exponential function, or none of these, the students can begin by inspecting consecutive differences, obtaining $\Delta p(1) = 14$, $\Delta p(2) = 38$, $\Delta p(3) = 74$, and $\Delta p(4) = 122$. Because the values of $\Delta p(x)$ are not constant, $p(x)$ cannot be a linear function. Because the values of $\Delta p(x)$ do not change in a linear fashion, $p(x)$ must not be a quadratic function. By examining the ratios $\dfrac{\Delta p(x)}{p(x)}$ the students discover that $\dfrac{14}{2} \neq \dfrac{38}{16} \neq \dfrac{74}{54} \neq \dfrac{122}{128}$. Because the ratios are not constant, the function $p(x)$ must also not be exponential.

Solutions for "Savvy Surveyor"

1. *a–c.* Students should draw a boundary through one of the vertices of △ABC and the midpoint of the opposite side, as shown on the right. They might start at different vertices, but their boundary should approximate a median of the triangle. If a boundary does not pass through a vertex, then at least one of the resulting regions will not be a triangle. Such a boundary, if it is a straight line segment, will intersect two sides of the triangle but not the third, thus creating a triangular and a quadrilateral region.

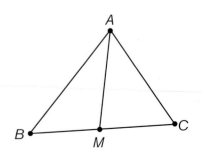

d. Both triangles ($\triangle ACM$ and $\triangle ABM$) formed by the median have base x and height h (see the figure on the right). Thus, their areas are equal.

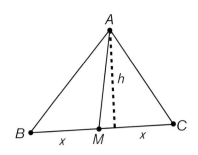

e. In each case, the student should note that the bases and heights, respectively, of the triangles are the same, and hence their areas are equal. If the students are using a dynamic geometry construction, they can also note that as they drag vertices of the original triangle, the area measurements of the two triangles created by a median remain the same.

2. *a.* One solution is shown on the right. The segments divide $\overline{BC}$ into three equal parts and hence divide $\triangle ABC$ into three triangles of equal area. Most of the solutions proposed by students will be like this, though they may start from a different vertex of $\triangle ABC$.

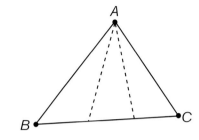

b. Answers will vary. Students constructing the centroid by using dynamic geometry software can measure the three regions and drag vertices to generate a compelling demonstration, but not a proof, of the proposition. Step 3 calls for a proof, which is given below.

3. *a–b.* Because *D*, *E*, and *F* are midpoints of the sides of $\triangle ABC$, by definition $\overline{GE}$ is a median of $\triangle ABG$; $\overline{GF}$ is a median of $\triangle ACG$; and $\overline{GD}$ is a median of $\triangle BCG$. Hence, from the results that the students obtained in step 1, it is clear that $a = b$, $c = d$, and $e = f$. But, since $\overline{AD}$ is a median of $\triangle ABC$, it is also clear that $\triangle ABD$ and $\triangle ACD$ have the same area. Therefore, $d + e + f = a + b + c$. Substituting the equalities above into this equation and simplifying yield the result $a = f = e = b$. Similarly, since $\overline{CE}$ is a median of $\triangle ABC$, it is clear that $\triangle ACE$ and $\triangle BCE$ have the same area. Therefore, $c + d + e = a + b + f$. Substituting the equalities already established and simplifying again lead to the result $d = a$. Thus, $a = b = c = d = e = f$.

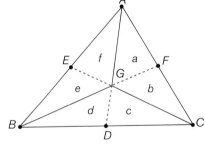

4. *a–b.* Students should note that the two triangles are not congruent. However, they should observe that the triangles have two pairs of sides that are equal in length, and the included angles of the equal sides are supplementary. The students then examine the upper figure on the right, in which $\angle ABC$ and $\angle DEF$ are supplementary, $AB = DE$, and $BC = EF$. Hence, if $\overline{FE}$ is extended to $\overline{FC'}$ so that $FE = EC'$, as in the lower figure, then $\triangle ABC \cong \triangle DEC'$ by side-angle-side, and $\overline{DE}$ is a median of FDC'. Hence, $\triangle DEF$ and $\triangle DEC'$ have the same area. This proves that $\triangle DEF$ and $\triangle ABC$ have the

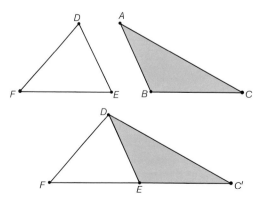

same area. Most proofs given by the students will be less formal and will involve such statements as the following: "The two triangles can be joined along segments $\overline{DE}$ and $\overline{AB}$ to form a single triangle with median $\overline{DE}$ separating the two original triangles. The two original triangles must therefore have the same area."

c–d. All the students' generalizations should express essentially the same idea: "If two sides of one triangle are congruent, respectively, with two sides of another triangle, and their included angles are supplementary, then the two triangles have the same area."

Solutions for "A Hex on Pythagoras"

1–3. If students list all the areas of the squares and triangles in all the Pythagorean hexagons that they create, then they should notice that the areas of the four triangles are equal in every case. Similarly, if they look at the squares of the lengths of the segments in the figures, they should discern—aided by their previous experience with the Pythagorean theorem—that the sum of the areas of the two smaller squares equals the area of the larger square. If they look at the sum of the squares of the segments along the perimeter of the Pythagorean hexagon, they may discover that this sum is four times the sum of the squares of the sides of $\triangle ABC$ and eight times the square of the hypotenuse of $\triangle ABC$.

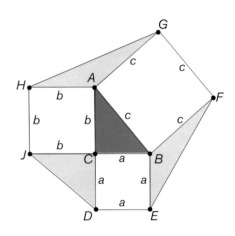

4. There is more than one way to prove the conjecture, "In any Pythagorean hexagon, the areas of the four triangles are equal." The simplest proof applies the fact that the students explored in Savvy Surveyor and summarized in their solution to step 4(*c*) in that activity. In the hexagon above, $AB = BF$, $CB = BE$, and $\angle CBA$ and $\angle EBF$ are supplementary. Therefore, $\triangle ABC$ has the same area as $\triangle EBF$. The students can make the same argument to show that $\triangle ABC$ has the same area as $\triangle HAG$ and $\triangle DCJ$.

5–7. Nothing in the preceding proof assumes that $\triangle ABC$ is a right triangle. The proof depends only on the result from Savvy Surveyor — a result that was true for all triangles, not just right triangles. Altering the construction of the Pythagorean hexagon so that $\triangle ABC$ is not a right triangle and then dragging the vertices to update the measurements and calculations leads to the discovery that the areas of the triangles remain the same as that of $\triangle ABC$, suggesting that the result is true regardless of whether $\triangle ABC$ is a right triangle. Because the proof of the conjecture in step 4 did not depend on $\triangle ABC$ being a right triangle, the same proof works for the general case in which $\triangle ABC$ is not assumed to be a right triangle. Like the other problems posed in this book, this one presents a challenge that you can encourage your students to accept if you wish. It invites the students to reflect on the principle behind their

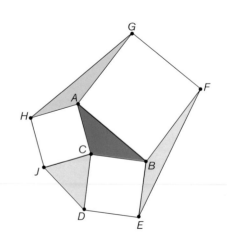

Navigating through Reasoning and Proof in Grades 9 through 12

proofs and to ask what other problems they can solve by applying the same principle. For example, they can quickly prove that the Pythagorean octagon generated by any convex quadrilateral (see the illustration on the right) has four triangles, the sum of whose areas is twice the area of the original quadrilateral. Furthermore, the sum of the areas of opposite triangles in the octagon equals the area of the quadrilateral.

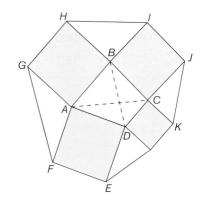

Solutions for "Centroid of Gravity"

1. *a–b.* The figure on the right shows nine points that approximately satisfy the criterion for being on the locus. These points appear to fall roughly on a circle centered at the midpoint of $\overline{QR}$. Students should conjecture that the locus of points is a circle centered at the midpoint of $\overline{QR}$ with a diameter of length QR.

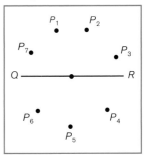

2. *a–b.* Students construct the circle centered at the midpoint of $\overline{QR}$ with a diameter of length QR. Those who are working with dynamic geometric software can then drag a point P on the circle while updating measurements and calculations. In the process, these students should see that the sum of the squares of PQ and PR remains constant at QR^2. If they choose a point N or N' in the exterior or the interior of the circle, respectively (see the figure to the right), they will discover that the sum of their squared distances from Q and R does not equal QR^2. Thus, their results from the dynamic geometric construction should suggest that the locus

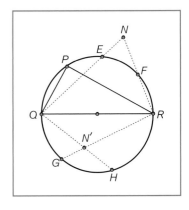

of points satisfying the criterion is the circle centered at the midpoint of $\overline{QR}$ with a diameter of length QR. To prove that this is so, as they are asked to do in (*b*), they can refer to a figure like that above. By the inscribed angle theorem, $\angle QPR$ is a right angle for points P on the circle other than Q and R. Hence, $\triangle QPR$ is a right triangle, and the Pythagorean theorem guarantees that $PQ^2 + PR^2 = QR^2$. Therefore, P belongs to the locus of points. Both Q and R also belong to the locus of points, since $QQ^2 + QR^2 = QR^2$ and $QR^2 + RR^2 = QR^2$. The secant-secant theorem ensures that for points on the exterior of the circle, such as N, $\angle QNR$ is equal to $\dfrac{m(\text{arc } QGR) - m(\text{arc } EF)}{2}$. However, this expression is equal to $90 - \dfrac{m(\text{arc } EF)}{2}$, which is less than 90°. Therefore, $\triangle QNR$ is not a right triangle, and by the converse of the Pythagorean theorem, $NQ^2 + NR^2$ does not equal QR^2. Thus, N does not belong to the locus. In a similar manner, the chord-chord and secant-tangent theorems for angles eliminate all points on the exterior and the interior of the circle. Therefore, the locus of points is exactly the set of points on the circle.

3. *a–b*. The figure below shows the construction. $QR = 2$, and $(QR)^2 = 4$. The equation of the locus of points (x, y) such that the squared distance of (x, y) to $(-1, 0)$ plus the squared distance of (x, y) to $(1, 0)$ is $(QR)^2$ is $x^2 + y^2 = 1$.

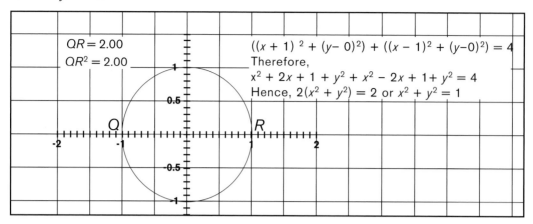

c. The equation of the unit circle is $x^2 + y^2 = 1$. But this is equivalent to the distance criterion specified for the locus of points, as seen in the figure above. Therefore, a point (x, y) is on the unit circle if and only if it satisfies the criterion for being in the locus of points.

4. *a*. For fixed values of $K > QR^2 = 4$, the locus of points is still a circle centered at the origin, the midpoint of $\overline{QR}$, which is no longer a diameter but in the interior of the circle, whose the diameter is $\sqrt{2(K-2)}$.

$$\left(x - (-1)\right)^2 + \left(y - 0\right)^2 + \left(x - 1\right)^2 + \left(y - 0\right)^2 = K$$

$$x^2 + 2x + 1 + y^2 + x^2 - 2x + 1 + y^2 = K$$

$$2x^2 + 2y^2 + 2 = K$$

$$x^2 + y^2 = \frac{K - 2}{2}$$

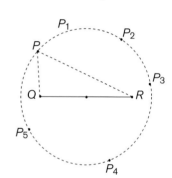

Sample locus of points P_k for which $(PQ)^2 + (PR)^2 = K > (OR)^2$

b. If $K = 2$, then only the origin can be in the locus of points. If K is less than 2, then the triangle inequality theorem guarantees that no point *off* the segment $\overline{QR}$ is in the locus of points, and the betweenness property ensures that no point *on* $\overline{QR}$ is in the locus of points. Therefore, the locus is the empty set. (The triangle inequality theorem holds that the sum of the lengths of any two sides of a triangle is greater than the length of the third side, and the betweenness property states that if a point C is between A and B and on segment AB, then $AC + CB = AB$.) If $2 < K < 4 = (QR)^2$, then the locus of points is still a circle centered at the origin, the midpoint of $\overline{QR}$, but Q and R and many other points on $\overline{QR}$ lie in the exterior of the circle. The diameter of the circle is again $\sqrt{2(K-2)}$.

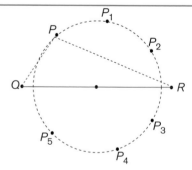

Sample locus of points P_k for which $(PQ)^2 + (PR)^2 = K < (OR)^2$

c–d. Suppose $K = 10$. Then the locus of points (x, y) such that the squared distance from (x, y) to $Q(-1, 0)$ plus the squared distance from (x, y) to $R(1, 0)$ equals 10 is the set of points (x, y) satisfying the equation

$$\left[(x - (-1))^2 + (y - 0)^2\right] + \left[(x - 1)^2 + (y - 0)^2\right] = 10.$$

This equation simplifies algebraically to the equation

$$x^2 + y^2 = \frac{10 - 2}{2}.$$

This is the equation of the circle of radius 2 centered at the origin. Its diameter is $\sqrt{2(10 - 2)}$, as predicted above. If K is changed to any value greater than 2, the result is the same except that the equation is $x^2 + y^2 = \dfrac{K - 2}{2}$, which is the equation of a circle centered at the origin with radius $\sqrt{\dfrac{K - 2}{2}}$. This result is merely one instance of a far more general result that your students could discover by using dynamic geometry software and progressively modifying the assumptions of the problem, though they probably could not prove this result. A proof of the generalization of the result in step 4 follows:

Proposition: If (x_1, y_1), (x_2, y_2), ..., (x_n, y_n) is any set of points and $\bar{x} = \dfrac{x_1 + x_2 + \cdots + x_n}{n}$ and $\bar{y} = \dfrac{y_1 + y_2 + \cdots + y_n}{n}$, then the locus of points P such that the sum of the squared distances from P to the points (x_1, y_1), (x_2, y_2), ..., (x_n, y_n) equals a fixed value K is a circle centered at $(\bar{x}, \bar{y})$, or is the empty set, or is simply the point $(\bar{x}, \bar{y})$.

Proof: To be in the locus of points, (x, y) must satisfy the following criterion:

$$\sum_{i=1}^{n}\left[\left(x - x_i\right)^2 + \left(y - y_i\right)^2\right] = K$$

$$\therefore \sum_{i=1}^{n}\left[x^2 - 2xx_i + x_i^2 + y^2 - 2yy_i + y_i^2\right] = K$$

$$\therefore nx^2 - 2x\sum_{i=1}^{n}x_i + ny^2 - 2y\sum_{i=1}^{n}y_i = K - \sum_{i=1}^{n}x_i^2 - \sum_{i=1}^{n}y_i^2 = H \quad \left(\text{some constant}\right)$$

Dividing by n yields $x^2 - 2x\dfrac{\sum_{i=1}^{n}x_i}{n} + y^2 - 2y\dfrac{\sum_{i=1}^{n}y_i}{n} = \dfrac{H}{n}$.

$$\therefore \quad x^2 - 2x\bar{x} + y^2 - 2y\bar{y} = \frac{H}{n}.$$

Adding $\bar{x}^2$ and $\bar{y}^2$ to both sides gives $x^2 - 2x\bar{x} + \bar{x}^2 + y^2 - 2y\bar{y} + \bar{y}^2 = \dfrac{H}{n} + \bar{x}^2 + \bar{y}^2$.

$$\therefore \left(x - \bar{x}\right)^2 + \left(y - \bar{y}\right)^2 = C \quad \left(\text{some constant}\right)$$

If C is negative, the locus of points is the empty set. If $C = 0$, the locus of points is the centroid $(\bar{x}, \bar{y})$. If C is positive, the locus is a circle.

Solutions for "Binomial Bonanza"

Don't Lose Your Heads – Part 1

1. Students' ideas will vary about the number of tosses — 10, 20, or 40 — that would give Kim her best chance of winning. The purpose of this step is to motivate the students to commit to a theory that they will want to test.

The sample solutions that appear below for steps 2–9 are based on the following simulation results for 50 repetitions of each version of the game. Each result represents the proportion of heads in one repetition (trial) of the simulation.

10-Toss Game	
.4	.5
.1	.3
.5	.5
.4	.5
.5	.5
.8	.6
.2	.7
.7	.6
.4	.7
.7	.5
.6	.7
.8	.4
.6	.5
.5	.3
.4	.6
.7	.5
.3	.5
.5	.4
.7	.6
.4	.5
.6	.6
.3	.5
.3	.3
.7	.6
.7	.6

20-Toss Game	
.5	.55
.6	.4
.75	.45
.6	.35
.55	.4
.5	.6
.35	.5
.55	.45
.4	.45
.4	.45
.45	.65
.6	.35
.5	.6
.55	.3
.65	.3
.6	.6
.6	.75
.4	.5
.5	.45
.5	.5
.5	.45
.5	.55
.4	.5
.5	.45
.35	.5

40-Toss Game	
.475	.425
.425	.475
.55	.425
.525	.625
.375	.5
.6	.475
.475	.45
.35	.625
.475	.375
.525	.65
.425	.625
.35	.525
.65	.4
.55	.475
.5	.575
.525	.375
.5	.45
.525	.5
.475	.45
.6	.55
.475	.45
.45	.65
.475	.425
.5	.45
.45	.375

2. Histograms for the sample data follow:

A Histogram of 50 *p*-Values for the 10-Toss Game

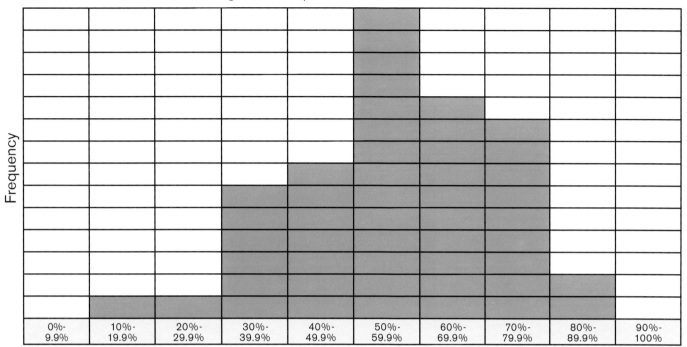

A Histogram of 50 *p*-Values for the 20-Toss Game

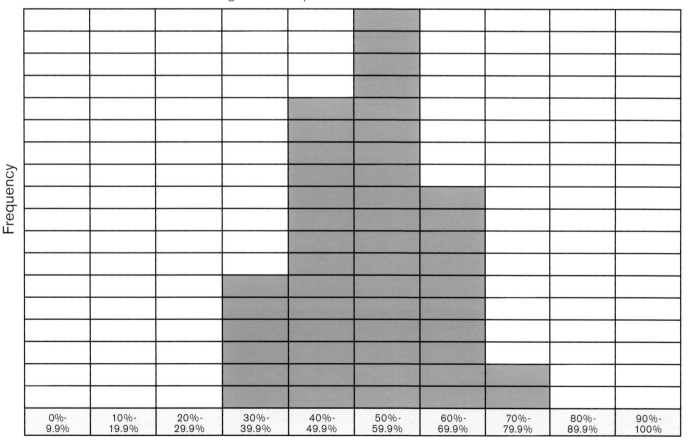

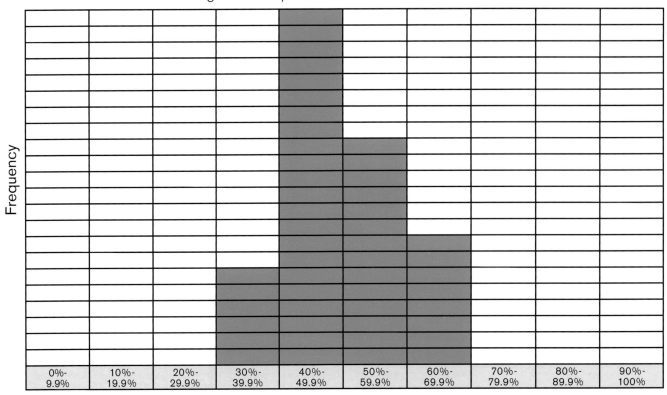

A Histogram of 50 *p*-Values for the 40-Toss Game

0%-9.9%	10%-19.9%	20%-29.9%	30%-39.9%	40%-49.9%	50%-59.9%	60%-69.9%	70%-79.9%	80%-89.9%	90%-100%

3. The following box-and-whiskers plots summarize the results of the simulations of the three versions of the game.

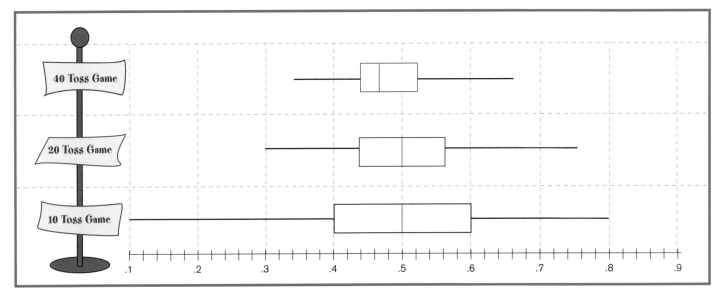

4–6. The sample data clearly indicate that the 10-toss game would be the best of the three possibilities for Kim. She won 19 of the simulated 10-toss games, 10 of the 20-toss games, and 12 of the 40-toss games. The means of the three distributions are .516, .497, and .491, respectively, for the 10-, 20-, and 40-toss games. These results suggest that as the number of tosses increases, the average number of heads on each trial stays approximately constant at 50 percent heads. It is clear that the spread in the distributions, for both the histograms and the box plots, is shrinking as the number of tosses increases. This pattern suggests that as the number of tosses grows, the variation from the means of 50 percent will be smaller. In fact, the standard deviations of the 10-toss, 20-toss, and 40-toss games are, respectively, .154, .101, and .079.

7. If the spreads of the distributions continue their rapid decline as the number of tosses in a game increases, a point will come at which it is virtually certain that Brenda will win, since virtually all the proportions of heads resulting from the trials will fall between 40 percent and 60 percent. It is interesting to note that in the sample simulation data, Kim won more 40-toss games (12 games out of 50) than she won 20-toss games (10 out of 50). This result is not very typical, and there is a high probability that a comparison of it with the results of other simulations would lead to the conjecture that it is likely, though not certain, that Kim will lose more games if she plays the 40-toss version than if she plays the 20-toss version.

8. Possible evaluations of the claims of Kim's friends, on the basis of the sample data, follow:

- Andrea claims that Kim should require Brenda to toss the coin 40 times instead of 20. Andrea's claim appears dubious, since doubling the number of tosses did not seem to improve Kim's results very much, whereas going to the 10-toss game almost doubled the number of her wins. Furthermore, the spread of the data shrinks significantly around the central tendency of 50 percent as the number of tosses increases. This evidence suggests that Kim's chances would be better if Brenda tossed the coin fewer times in a game.

- Tracy claims that the more times Brenda tosses the coin, the closer the number of heads will be to exactly 50 percent, so she contends that tossing the coin 40 times instead of 20 would make Brenda more likely to win. The data do not support Tracy's claim. In the 10-toss simulation of the game, 14 of the 50 trials ended up with exactly 50 percent heads. In the 20-toss simulation, 13 of the 50 trials ended up with exactly 50 percent heads. In the 40-toss simulation, only 5 of the 50 trials ended up with exactly 50 percent heads. On the basis of the data, it appears that the chances of getting exactly 50 percent heads actually goes down as the number of tosses goes up. This makes sense, since with 40 tosses, the chances of getting 18, 19, 20, 21, or 22 heads are almost the same, which means that none of them can have a probability much larger than 20 percent. If one tosses the coin 1000 times, the chances of getting 495, 496, 497, 498, 499, 500, 501, 502, 503, 504 or 505 heads are almost the same, which means that none of these outcomes can have a probability much larger than 10 percent. Thus, as the number of tosses increases, the number of outcomes that are virtually as frequent as 50 percent heads grows, thereby making the probability of exactly 50 percent heads shrink.

- Thia claims that Kim's chances of winning are better if Brenda tosses the coin 10 times instead of 40, arguing that the percentage of heads gets closer to 50 percent as the number of tosses increases. Although it is true that among her three choices (10-toss game, 20-toss game, 40-toss game) Kim has her best chance of winning with the 10-toss game, Thia's general claim — that the percentage of heads gets closer to 50 percent as the number of tosses increases — is not supported by the sample data. Fourteen of the 10-toss trials resulted in exactly 50 percent heads. You can't get closer than that by tossing the coin more times!

- Brenda claims that since she wins only if she gets 8, 9, 10, 11, or 12 heads, Kim has a lot more chances of winning than she does. Brenda's claim is not supported by the sample data. Even though more of the possible outcomes of the 20-toss experiment would give Kim the win, most of these possible outcomes are very unlikely. In the 20-toss simulation, the outcomes that make Brenda the winner — 8, 9, 10, 11, and 12 heads — occurred in 40 out of the 50 trials.

9. The students' explanations of why the probability of getting heads on a toss of a fair coin is said to be one-half will vary. Students should start to look for an explanation based on an assumption of the equal likelihood of heads and tails. To assume that the coin is "fair" is essentially to suppose that heads and tails are equally probable. Therefore, the probability of heads must be one-half, and the probability of tails must also be one-half, since the sum of the probabilities of the outcomes in the sample space is 1.

Solutions for "Pascal's Bonanza"

Don't Lose Your Heads – Part 2

1. The assumptions in the experiment of tossing a fair coin dictate that the probability of heads is one-half. To assume that the coin is "fair" is essentially to say that heads and tails are equally probable. Therefore, the probability of heads must be one-half, and the probability of tails must be one-half, since the sum of the probabilities of the outcomes in the sample space is assumed to be equal to 1.

2. *a.* Students' descriptions of how the tree diagram represents the experiment and outcomes of tossing a coin two times will vary, but all students should observe that the tree diagram breaks the experiment into stages. They should note that for each outcome of the first stage (the first toss), there are two possible outcomes in the second stage. Thus, they should see that the total number of possible outcomes is the number of branches at the right end of the tree, and each branch is associated with a different sequence of heads and tails in the two stages.

   b. Students' explanations of why the probability of each outcome in the sample space of the experiment of tossing a coin two times is $\frac{1}{4}$ will vary. Most students will focus on the tree diagram and assert that at each stage (toss), each path is equally likely, and, therefore, the two-stage paths are all equally likely. Regardless of the outcome on the first toss, the probabilities of heads or tails on the second toss are equal. The area model shown below captures these intuitive explanations.

		Outcome of Second Toss	
		H ($p = .5$)	T ($p = .5$)
Outcome of Frist Toss	H ($p = .5$)	HH ($p = .25$)	HT ($p = .25$)
	T ($p = .5$)	TH ($p = .25$)	TT ($p = .25$)

   c. In a two-toss experiment with a fair coin, the probability that the number of heads equals the number of tails, or *P*(HT or TH), is $\frac{1}{4} + \frac{1}{4}$, or $\frac{1}{2}$.

3. *a.* A tree diagram for a three-toss experiment with a fair coin appears at the top of the next page. Its data are translated into a list of the outcomes in the sample space, and the list is in turn translated into a probability table for the outcomes.

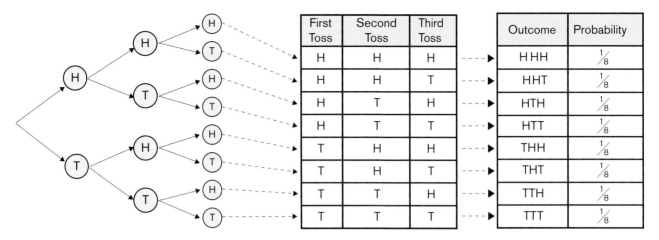

First Toss	Second Toss	Third Toss		Outcome	Probability
H	H	H		HHH	$\frac{1}{8}$
H	H	T		HHT	$\frac{1}{8}$
H	T	H		HTH	$\frac{1}{8}$
H	T	T		HTT	$\frac{1}{8}$
T	H	H		THH	$\frac{1}{8}$
T	H	T		THT	$\frac{1}{8}$
T	T	H		TTH	$\frac{1}{8}$
T	T	T		TTT	$\frac{1}{8}$

Because each of the possible outcomes of the 2-toss experiment are equally likely and, regardless of the results of 2 tosses, the probability of heads and tails are equal on the third toss, then the probability of each of the eight branches in the 3-toss game are equally likely. Therefore, the probability of each outcome is $\frac{1}{8}$. The following area model shows this as well:

Outcome of Third Toss

Outcomes of First Two Tosses

	H ($p = .5$)	T ($p = .5$)
HH ($p = .25$)	HHH ($p = .125$)	HHT ($p = .125$)
HT ($p = .25$)	HTH ($p = .125$)	HTT ($p = .125$)
TH ($p = .25$)	THH ($p = .125$)	THT ($p = .125$)
TT ($p = .25$)	TTH ($p = .125$)	TTT ($p = .125$)

b. The completed table follows:

Event: Total number of heads	0 heads	1 head	2 heads	3 heads
Number of outcomes	1	3	3	1

4. *a–b.* The students should observe that the number of outcomes doubles from stage to stage. From the tree diagram they should see that each branch of the $(N-1)$-toss game determines two branches of the N-toss game. Thus, there are twice as many branches in the N-toss game as in the $(N-1)$-toss game. Because there are 2 outcomes at the first stage and the number doubles from stage to stage, there are $(N-1)$ doublings from the first stage to the Nth stage. Hence, there are $2 \times 2^{N-1} = 2^N$ outcomes at the Nth stage. This argument can be expressed by using mathematical induction.

5.

1st Toss	2nd Toss	3rd Toss	4th Toss	Probability of Outcome
H	H	H	H	$\frac{1}{16}$
H	H	H	T	$\frac{1}{16}$
H	H	T	H	$\frac{1}{16}$
H	H	T	T	$\frac{1}{16}$
H	T	H	H	$\frac{1}{16}$
H	T	H	T	$\frac{1}{16}$
H	T	T	H	$\frac{1}{16}$
H	T	T	T	$\frac{1}{16}$
T	H	H	H	$\frac{1}{16}$
T	H	H	T	$\frac{1}{16}$
T	H	T	H	$\frac{1}{16}$
T	H	T	T	$\frac{1}{16}$
T	T	H	H	$\frac{1}{16}$
T	T	H	T	$\frac{1}{16}$
T	T	T	H	$\frac{1}{16}$
T	T	T	T	$\frac{1}{16}$

a. In the 4-toss game there are 16 possible outcomes. The pattern in the list consists of the halving of the alternating blocks of heads and tails from one column to the next. The first column contains a sequence of eight heads followed by eight tails. This corresponds to the fact that the two nodes for the first toss in the tree diagram each lead to eight different branches. The column for the second toss has alternating blocks of four heads and four tails. Again, this corresponds to the fact that in the tree diagram the four nodes of the second toss each lead to four different branches. The column for the third toss has alternating blocks of two heads and two tails. And in the column for the fourth toss, the size of the alternating blocks of heads and tails is 1. In general, in a listing of the possible outcomes of the N-toss game, the size of the alternating block of heads and tails in the kth column is 2^{N-k}.

b. The probability that exactly half of the tosses turn out heads in the 4-toss game is the sum of the probabilities of the outcomes with exactly two heads (see the shaded rows in the table above). There are six such outcomes, so the probability of the event "Exactly half of the tosses are heads" is $6 \times \frac{1}{16}$, or $\frac{6}{16}$, or $\frac{3}{8}$. Note that this is less than the probability of getting exactly half heads in the 2-toss game (see the solution for step 2(c). The probability of getting exactly 50 percent heads decreases as the number of tosses increases.

6. *a–d.* The students can observe many patterns to help them fill in the row in the table for the 4-toss experiment. The simplest pattern (and the most effective one to use) is that after the first row, the entry in each cell is the sum of the entries in the cell directly above it and the first cell above and to the left of it. All cells in the first column have a 1, which corresponds to the fact that in the *N*-toss game there is only one possible outcome that has all tails and no heads (or all heads and no tails). The completed table is shown below.

Event: Total Number of Heads

	0	1	2	3	4	5	6	7	8	9	10
1-toss	1	1	0	0	0	0	0	0	0	0	0
2-toss	1	2	1	0	0	0	0	0	0	0	0
3-toss	1	3	3	1	0	0	0	0	0	0	0
4-toss	1	4	6	4	1	0	0	0	0	0	0
5-toss	1	5	10	10	5	1	0	0	0	0	0
6-toss	1	6	15	20	15	6	1	0	0	0	0
7-toss	1	7	21	35	35	21	7	1	0	0	0
8-toss	1	8	28	56	70	56	28	8	1	0	0
9-toss	1	9	36	84	126	126	84	36	9	1	0
10-toss	1	10	45	120	210	252	210	120	45	10	1

(Experiment — row labels at left)

7. *a.* In a 10-toss game, the outcomes corresponding to 4, 5, or 6 heads would lead to wins for Brenda. A 10-toss game would have 2^{10}, or 1024, possible outcomes. There are 210 + 252 + 210, or 672, outcomes corresponding to 4, 5, or 6 heads. So in 672 out of 1024 outcomes, Brenda would in theory win.

   b. Winning in 672 out of 1024 outcomes would mean that the theoretical probability of Brenda winning the 10-toss game is .66.

   c. In the sample data reported above, Brenda won 31 of the 50 trials of the simulated 10-toss game. Thus, on the basis of the data, the experimental probability of Brenda winning was .62. This value is reasonably close to .66, suggesting that the 50-trial simulation gave a good picture of the theoretical probabilities.

8. For the 20-toss game:

   a. There are 2^{20}, or 1,048,576, possible outcomes. The outcomes corresponding to 8, 9, 10, 11, or 12 heads would lead to wins for Brenda.

   b. From the entries generated for the 20th row of Pascal's triangle, the students can determine that in 125,970 + 167,960 + 184,756 + 167,960 +125,970 out of 1,048,576 possible outcomes Brenda would in theory win. This means that there is a theoretical probability of Brenda winning the 20-toss game is .74.

    c. In the sample data reported above, Brenda won 40 out of 50 trials of the simulated 20-toss game. Thus, on the basis of the simulation, the experimental probability of Brenda winning was .80.

For the 40-toss game:

    a. There are 2^{40} possible outcomes. The outcomes corresponding to 16, 17, 18, 19, 20, 21, 22, 23, or 24 heads would lead to wins for Brenda.

    b. From the entries generated for the 40th row of Pascal's triangle, the students can determine that the total number of these outcomes is 930,340,830,120. This corresponds to a theoretical probability of .85 that Brenda would win.

    c. In the sample data reported above, Brenda won 38 out of 50 trials of the simulated 40-toss game. Thus, on the basis of the simulation, the experimental probability of Brenda winning was .76.

9. *a.* Kim was justified in not wanting to play the game with Brenda! From a theoretical standpoint, the game is very unfair — even the 10-toss version.

    b. The 50-trial simulations are likely to have given a sufficiently accurate picture of the theoretical probabilities for the students to conclude with a high degree of certainty that the game was unfair to Kim.

Solutions for "Derek Jeter Steps into the Box"

1–2. Students' estimates of the average number of boxes will vary, as will their explanations of their reasoning. Some students may say that a family of Derek Jeter fans would need to buy about 2.5 boxes, offering an argument such as the following: "It's almost certain that you will get a Derek Jeter card by the time you have bought 4 boxes of cereal. So the average should be somewhere around 2 boxes."

3. Students' statements about the assumptions in their estimates will vary, but the students should see that a solution to the problem assumes that boxes of Sweet & Soggy Squares are distributed randomly to each of the stores that sell the cereal and that the purchase of one box doesn't affect the probabilities of the outcomes from purchasing other boxes.

4–5. To test their conjectures, the students could use various methods. Step 5 suggests that they simulate the situation of obtaining a Derek Jeter card in a box of Sweet & Soggy Squares by randomly generating integers from 1 to 4 with equal likelihood and stopping when they get a the integer that represents the Derek Jeter card — a 2 in the simulation outlined in the activity. Sample results from 30 trials of this experiment are listed in the following chart, and the frequencies of these results are tabulated in the frequency table that appears at the top of page 182.

Number of Trial	List of Integers																			Length of Trial
1	3	1	4	2																4
2	3	4	2																	3
3	2																			1
4	3	3	4	3	3	4	2													7
5	4	3	3	3	4	3	2													7
6	1	2																		2
7	2																			1
8	2																			1
9	2																			1
10	3	2																		2
11	3	4	1	3	1	4	3	2												8
12	2																			1
13	4	4	4	2																4
14	1	4	2																	3
15	3	1	2																	3
16	1	2																		2
17	3	4	1	2																4
18	1	1	3	4	1	1	3	4	1	1	3	1	1	3	4	1	4	4	2	19
19	2																			1
20	4	1	1	2																4
21	3	3	4	2																4
22	3	2																		2
23	4	3	4	1	2															5
24	3	1	4	3	1	3	1	4	2											9
25	4	2																		2
26	1	2																		2
27	2																			1
28	2																			1
29	2																			1
30	3	2																		2

A frequency table follows for the sample results from 30 trials, showing the number of trials resulting in the lengths displayed in the table on the previous page.

Length of trial	1	2	3	4	5	6	7	8	9	10	11	12	13	14	15	16	17	18	19
Frequency	9	7	3	5	1	0	2	1	1	0	0	0	0	0	0	0	0	0	1

6. The data in the frequency table above yield 3.6 as the average length of a trial. This result would neither validate nor refute the sample estimate of 2.5 given in the solution to step 1. A student making that estimate could be very convinced that the average would be close to 2 and certainly less than 4. Students sometimes confound expected value with their intuitions about probability. The probability of getting the Derek Jeter card in three or fewer purchases is about 55 percent. The probability of getting the Derek Jeter card in two or fewer purchases is about 44 percent. The sample response of 2.5 would seem to reflect an intuitive idea that the average number of boxes needed would be the point where the probability is 50 percent — hence, somewhere around 2.5 boxes.

7. The following table shows a sample of pooled results from simulations of 30 trials of the experiment by each of six students.

Length of Trial / Frequency	1	2	3	4	5	6	7	8	9	10	11	12	13	14	15	16	17	18	19
Student 1	9	7	3	5	1	0	2	1	1	0	0	0	0	0	0	0	0	0	1
Student 2	4	11	5	1	2	1	1	2	0	0	0	2	0	1	0	0	0	0	0
Student 3	9	6	3	3	1	2	3	1	1	0	0	1	0	0	0	0	0	0	0
Student 4	9	9	3	1	0	2	1	1	1	0	1	0	0	1	0	0	0	1	0
Student 5	7	6	4	4	2	3	0	0	1	1	1	1	0	0	0	0	0	0	0
Student 6	7	5	3	6	2	1	1	0	2	2	0	0	0	0	1	0	0	0	0
	--	--	--	--	--	--	--	--	--	--	--	--	--	--	--	--	--	--	--
Column totals	45	44	21	20	8	9	8	5	6	3	2	4	0	2	1	0	0	1	1
Relative frequency	.25	.244	.117	.111	.044	.05	.044	.028	.033	.017	.011	.022	0	.011	.006	0	0	.006	.006

8. a–c. Students can relate the numerator in the ratio

$$\frac{\text{Sum of the lengths of the 30 trials}}{\text{30 trials}}$$

to the total number of boxes purchased in 30 trials. They should see that the denominator represents the number of Derek Jeter cards in the 30 trials and must equal 30, since each trial stops when a box of cereal contains a Derek Jeter card. Therefore, in the equation

$$\text{Average length of a trial} = \frac{\text{Sum of the lengths of the 30 trials}}{\text{30 trials}},$$

the average length of a trial is equal to the reciprocal of the relative frequency of the Derek Jeter card in the boxes purchased in the 30 trials.

d. The average length of a trial in the sample of pooled data displayed in the table in step 7 above is 3.9 boxes of cereal. The numerator in the ratio that gives this average is the total number of boxes purchased in the 180 trials, or 700 boxes of cereal. The denominator is the number of Derek Jeter cards obtained in the 180 trials, or 180 cards, since each trial stops when a box contains a Derek Jeter card. The relative frequency of a Derek Jeter card in the pooled sample is thus $\frac{180}{700}$, or $\frac{9}{35}$. The average length of a trial is equal to the reciprocal of the relative frequency, or $\frac{35}{9}$, or approximately 3.9 boxes of Sweet & Soggy Squares.

9. *a.* Students' responses will vary but should express an understanding of the fact that the reciprocal of the relative frequency of Derek Jeter cards in larger and larger numbers of trials will, in theory, approach 4 with high probability in the long run. This is so because the relative frequency of the Derek Jeter cards in the whole population of cereal boxes is 25 percent. Thus, in theory, the average length of a trial will become arbitrarily close to 4 boxes, with high probability, as the number of trials increases. Therefore, the theoretical average is 4 boxes.

b. The accuracy of the students' 30-trial and 180-trial simulations in estimating the average length of a trial will vary, as will their explanations of the accuracy of their results. In the sample of simulated results from 30 trials, the average length of a trial (number of boxes needed to get a Derek Jeter card) was 3.6, and in the sample of simulated results from 180 trials, it was 3.9. Theoretically, this average length of a trial should get closer to 4, with high probability, as the number of trials increases. The average from the sample of simulated results from 30 trials, 3.6, has a relative error of $\frac{4 - 3.6}{3.6}$, or approximately 11 percent, when 3.6 is used to estimate the theoretical average of 4. Some people would consider this amount of error to be rather high. The average number of boxes from the sample of simulated results from 180 trials, or 3.9, has a relative error of about 3 percent. For most purposes, this amount of error indicates that a result is acceptably accurate.

10. *a.* If the students make a histogram of the frequencies in their pooled results from 180 trials, the graph is likely to show an approximation to the geometric distribution curve with a skew to the right. The students can use the data from their pooled set of 180 trials to get fairly accurate experimental probabilities to answer a variety of questions, including those posed in this step. On the basis of the sample data, students would see that if someone purchased four boxes of Sweet & Soggy Squares, the probability that he or she would *not* get a Derek Jeter card is 1 − (.25 + .244 + .117 + .111), or .278, as compared with a theoretical probability of .75 × .75 × .75 × .75, or about .316. Working with the sample data, students would conclude that someone making just one trip to the store and buying enough boxes of Sweet & Soggy Squares to have a 90 percent probability of getting a Derek Jeter card should buy 9 boxes of the cereal, as compared with 8 boxes, as determined from theoretical probabilities. Students could add the relative frequencies from the sample data set for trial lengths of 1 box, 2 boxes, 3 boxes, and so on, all the way up to 9 boxes before obtaining a sum of 90 percent for the relative frequencies. Using theoretical probabilities and letting *n* represent the number of boxes, they could find *n*, the number of boxes, in the following sum:

$$.25 + (.75)(.25) + (.75)^2(.25) + \ldots + (.75)^{n-1}(.25) = .9$$

They would discover that n is very close to 8 boxes. They could also approach this question by solving for n in the equation $\left(\dfrac{3}{4}\right)^n = .1$, determining that $n \approx 8$, since $\left(\dfrac{3}{4}\right)^8 \approx 0.1001$.

b. The theoretical probability that someone gets his or her first Derek Jeter card in the fourth box of cereal that he or she purchases is $\dfrac{3}{4} \times \dfrac{3}{4} \times \dfrac{3}{4} \times \dfrac{1}{4}$, or $\dfrac{27}{256}$, or approximately .105. Students' comparisons of this theoretical value to the experimental probability obtained from their pooled data set will vary. In the sample data for 180 trials presented above, the experimental probability that the Derek Jeter card arrives on the fourth purchase is $\dfrac{20}{180}$, or $\dfrac{1}{9}$, or .111. This has a relative error of only 5 percent when compared with the theoretical probability: $\dfrac{.111-.105}{.111} = \dfrac{.006}{.111} = .05.$

Solutions for "Late Shift at the Orange Bowl"

1. a–b. The students will propose a variety of designs for simulations (see the discussion of step 1 on pp. 67–68).

2. a. The students in the class might agree on one of several effective simulations to use (see pp. 68–69). The sample solutions offered below use data generated in the simulation presented as the second alternative in the text. In a trial of this simulation, Tony's bus arrives at the Orange Bowl before Rhonda's bus when the command **20*rand** on a TI-84 calculator generates a value that is less than the value that the command **10+15*rand** generates. The command **rand** produces real numbers between 0 and 1 in a uniform distribution, and the command **20*rand** produces real numbers between 0 and 20 in a uniform distribution. In the simulation, these numbers represent Tony's arrival times between 11:00 and 11:20. The command **10+15*rand** produces real numbers between 10 and 25 in a uniform distribution. In the simulation, these numbers represent Rhonda's arrival times between 11:10 and 11:25. This approach assumes that as long as Tony arrives even a fraction of a second before Rhonda, he will succeed in greeting her as she steps off her bus. It also assumes that the arrival times of the buses are independent of each other.

   b. In 100 sample trials of the simulation, Tony arrived first in 84 of the trials. Thus, for the sample data, the experimental probability of the event "Tony's bus arrives before Rhonda's bus" is .84.

3. a. All the students' experimental probabilities, each determined from the results of 100 trials of the simulation, will not be the same, but most of them are likely to give numbers close to 84 out of 100 trials. Indeed, this was the case in the sample results shown in 3(b) below, from 25 students, with each student conducting 100 trials of the simulation. Because the arrival times of the buses are randomly dispersed over their respective arrival intervals, the students can expect variability in the results of their simulations.

   b. Students' estimates of the probability on the basis of their pooled results will vary. The results of 25 sets of 100 trials of the simulation follow. The values represent the number of times in 100 trials that Tony arrived first.

80	82	94	86	86
86	82	88	84	79
86	84	80	89	83
73	85	84	84	84
80	77	85	83	87

The mean of this set of values is 83.6, and its standard deviation is 4.2. Thus, an estimate of the theoretical probability that Tony arrives before Rhonda is .836. All but two of the 25 simulations produced results between 76 and 92, each of which is roughly 8 trials (or approximately two standard deviations) from 83.6. Therefore, students analyzing the sample data could be fairly confident that the theoretical probability falls in the interval .836 ± .08. Students working with the sample data could thus estimate the uncertainty in their prediction of an 83.6 percent success rate for Tony to be 8 percent.

4–5. To investigate the theoretical probability of Tony's bus arriving before Rhonda's, the students show the set of all the possible arrival times of the two buses as an area made up of ordered pairs on a grid. The set of all ordered pairs (T, R), where $0 \leq T \leq 20$ and $10 \leq R \leq 25$ represents the set of all possible combinations of arrival times — the sample space for the problem. The rectangular region outlined below represents this set of ordered pairs.

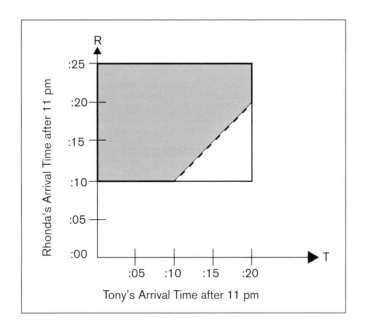

Next, the students consider the ordered pairs (T, R) in which $T < R$. These pairs represent the outcomes in which Tony's bus arrives before Rhonda's. The shaded area of the rectangle above represents this subset of all the ordered pairs (T, R). (The dotted segment that forms one side of this region represents the ordered pairs in which $T = R$.) The area of the rectangle is 20 × 15, or 300, square units. The area of the shaded subset of the rectangle is 300 square units minus the area of the unshaded triangle, or 300 − 50, or 250, square units. Thus, the theoretical probability that Tony arrives before Rhonda is $\frac{250}{300}$, or $\frac{5}{6}$, or $83\frac{1}{3}$ percent. This approach assumes that the probability of any subset of ordered pairs in the rectangular sample space is proportional to its area. In the context of probability, this is equivalent to assuming that the probability that Tony arrives in a time interval A is directly proportional to the length of the time interval A, where A is a subinterval of 11:00 –11:20 p.m., and that the probability that Rhonda arrives in a time interval B is directly proportional to the length of the time interval B, where B is a subinterval of 11:10–11:25 p.m.

6. Students' comparisons and contrasts of the theoretical probability with their experimental results will vary. In the solution for step 2(*b*), the result reported for the sample simulation of 100 trials provides a good estimate of the theoretical probability that Tony arrives first (Tony arrived before Rhonda in 84 of the trials). The solution for step 3 reports the pooled data from 25 sample simulations of 100 trials. Here some of the simulations consisting of 100 trials give estimates whose relative error is greater than 12 percent. However, when the data from the 25 simulations are combined to give data from 2500 trials, the relative error is less than one-half of one percent. Thus, for the sample data, the theoretical probability is well within the uncertainty interval estimated for the experimental probability. It would not be certain that the theoretical probability would be in that interval, but on the basis of the sample results, students could be very confident that it would be.

7. One example of a problem that students could solve by using the same geometric approach follows: "Assuming that the arrival intervals of Tony's and Rhonda's buses remain the same, find the probability that Tony and Rhonda arrive within 5 minutes of each other." The students would translate this into the mathematical statement $|T - R| \leq 5$, whose solution set is the region of the rectangle whose points are between the lines $T - R = 5$ and $T - R = -5$. This region is shaded below and has an area of 100 square units.

Thus, the probability that Tony and Rhonda arrive within 5 minutes of each other is $\frac{100}{300}$, or $33\frac{1}{3}$ percent.

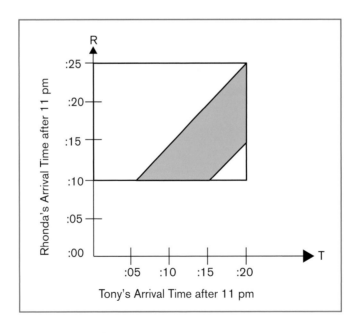

Solutions for "Pushing the Perimeter"

1. Students are likely to guess, incorrectly, that Kelsey's conjecture is true. Explanations might include the contention, "It seems logical." Or students might offer an incomplete or flawed statement, such as, "An increased perimeter means the length and width increase, which will increase the area."

2. You might decide to lead a group discussion about questions (*a*) and (*b*) before letting students investigate Kelsey's claim (in groups, pairs or individually).

   a. In deciding whether considering squares as a particular case of rectangles would help in evaluating Kelsey's conjecture, the students should review the definition of a square. Considering squares can lead quickly to the discovery that the original conjecture is false.

   b. In considering ways of organizing the charts to highlight patterns or trends, students might suggest keeping one dimension — either length or width — constant while varying the other. This organization could lead to interesting patterns but would not reveal much about the conjecture. Encourage your students to consider other arrangements of data, such as finding all rectangles with the same perimeter but different areas.

3. *a–c.* The students can observe many interesting patterns in their charts. They may have investigated rectangles whose perimeters are a larger number of linear units than the number of square units in their areas (for instance, a rectangle measuring 3 units by 5 units), and they may have investigated other rectangles (for example, a rectangle measuring 4 units by 5 units) whose areas are a larger number of square units than the number of linear units in their perimeters. Students should have little difficulty in finding two rectangles with the same perimeter but different areas (for example, a rectangle measuring 10 units by 10 units and a rectangle measuring 16 units by 4 units) and two rectangles with the same area but different perimeters (for example, a rectangle measuring 10 units by 10 units and a rectangle measuring 25 units by 4 units).

   d. Considering various rectangles with equal areas of 100 square units should lead students to the crux of the original conjecture. With some guidance, the students should realize that for a fixed area, assuming that the length and width are not restricted to whole numbers, the perimeter can become arbitrarily large. Consider the following argument: If x and y are the dimensions of a rectangle whose area is 100 square units, then $x \times y = 100$ square units. Since neither x nor y can be zero (otherwise, there would be no rectangle), the students can write $x = \dfrac{100}{y}$. As the value of y increases (and x decreases), the perimeter will become arbitrarily large. By contrast, the smallest perimeter of a rectangle with an area of 100 square units is 40 units: $4 \times \sqrt{100} = 40$.

4. Most students should now see that Kelsey's claim is false. A rectangle measuring 2 units by 13 units has a perimeter of 30 units and area of 26 square units. On the other hand, a rectangle measuring 7 units by 7 units has a smaller perimeter (28 units) but a larger area (49 square units). In other words, since the area and perimeter depend on two attributes — length and width — it is possible to make changes in the two attributes that result in different changes in area and perimeter.

5. In the two samples of work below, students justify the conclusion that Kelsey's conjecture is false.

Sample Response 1

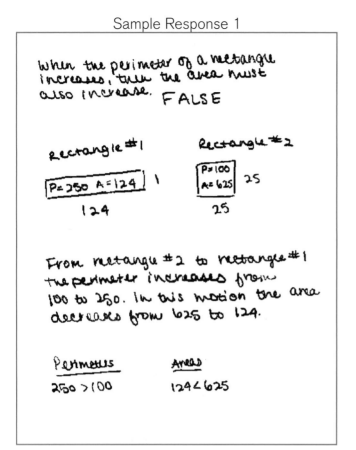

When the perimeter of a rectangle increases, then the area must also increase. FALSE

Rectangle #1   Rectangle #2

P=250 A=124   P=100
                      A=625   25
    124   25

From rectangle #2 to rectangle #1 the perimeter increases from 100 to 250. In this motion the area decreases from 625 to 124.

Perimeters   Areas
250 > 100   124 < 625

Sample Response 2

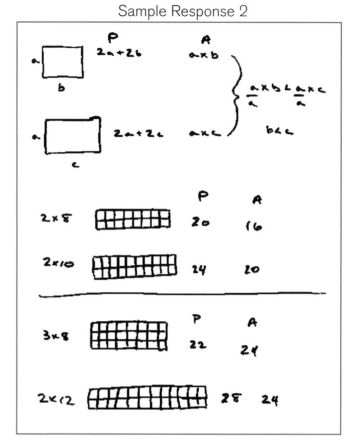

6. Kelsey's conjecture is true for circles and regular polygons. Explanations for both cases are provided in the text (see p. 78).

Solutions for "Heavy Metal at Maximum Volume"

Maximizing the Measurement — Part 1

1. Students may initially guess that cutting out squares of a smaller size will give a box of larger volume. They might reason that a large surface area accompanies a large volume. Alternatively, some might conjecture that a taller box, made by removing larger squares, can have a larger volume. The students' work with paper or tagboard is thus likely to yield different models, and the students' comparisons of their models will yield different ideas.

2. The students' new predictions of the size of the square to remove to create the box of maximum volume will vary, depending on their models and comparisons.

3. The students should use a computer with spreadsheet software or a graphing calculator to create tables. On a graphing calculator such as the TI-83 Plus, they might have access to a cell sheet application that behaves like a spreadsheet. In any case, by using the function editor and letting x stand for the length of the side of the cut-out square, the students can represent columns 2–5, respectively, by the functions **Y1(x)=16−2x, Y2(x)=30−2x, Y3(x)=x**, and **Y4(x)=x(16-2x)(30-2x)**. They can create the table from these functions by using the table-set (**TBLSET**) and table (**TABLE**) options on the calculator. **TBLSET**

allows students to specify the starting value for *x* and the increment in the *x*-value and updates the functions tables for **Y1** through **Y4** over the domain of the set. If students simply want to investigate the relative maximum of the volume function, then they should enter it into **Y1** and use the table features to explore the value of the maximum. Below are two screen images produced by a TI-83 calculator during such an investigation.

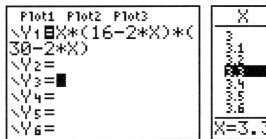 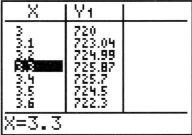

Volume function and resulting table for **TblStart=3** and **ΔTbl=.1**

a. The length of one side of the cut-out square must be greater than 0 centimeters and less than 8 centimeters to create a box.

b. Students' choices of the size of the increment will vary. It makes sense to start with a larger increment and use it to determine an interval for further exploration. Once students have found an interval with a potential maximum value, they should use a smaller increment to explore that interval. By repeating this process several times, they can get as close to the actual maximum as the precision of the calculator or spreadsheet will allow.

c. Column 2 (which shows the length of the resulting box) is the original length of the sheet metal (16 cm) less the lengths of two squares cut from each end:

$$\text{(value in column 2)}=16-2*\text{(value in column 1)}.$$

Column 3 (which shows the width of the resulting box) is the original width of the sheet metal (30 cm) less the widths of two squares cut from each end:

$$\text{(value in column 3)}=30-2*\text{(value in column 1)}.$$

Column 4 (which shows the height of the resulting box) is the length of the side of the square that is cut from the original rectangle:

$$\text{(value in column 4)}=\text{(value in column 1)}.$$

If the cut-out square has side *x*, then the length of the resulting box is $16 - 2x$, its width is $30 - 2x$, and its height is *x*. The students need to subtract $2x$ from the length and width since they are cutting a square from each end. The diagram below illustrates the situation.

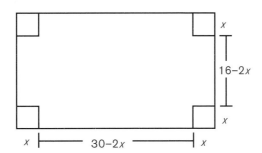

d. The volume of the box — the value in column 5 — is the product of the length, width, and height of the box:

(value in column 5)=
(16−2∗(value in column 1))∗(30−2∗(value in column 1))∗(value in column 1).

4. *a.* In column 5 in the table, the value for the volume, $V(x)$, of the box in terms of x is $x(16 − 2x)(30 − 2x)$.

   b. The graph below shows the volume function $V(x) = x(16 − 2x)(30 − 2x)$ in the interval $1 \leq x \leq 8$.

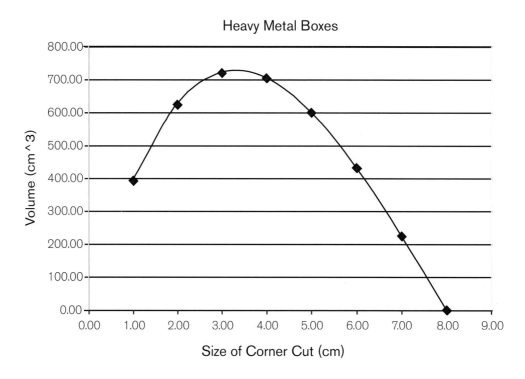

Heavy Metal Boxes

c. The *y*-value of the "high point" of the graph indicates the maximum volume.

d. By inspecting the graph, the students should be able to estimate a maximum of approximately 700 cm³. Using the "**trace**" function of the graphing calculator will permit the students to obtain a better estimate of the maximum point (for example, 730 cm³).

5. *a.* Three sample tables follow that display data for successively smaller ranges and increments. The graph shown above in the solution for step 4(*b*) displays the data in the first table below; graphs for the data in the second and third tables are shown below the respective tables.

Starting with x = 1 and increasing in increments of 1 cm

Side Length of Cut-Out Square (cm)	Length of Resulting Box (cm)	Width of Resulting Box (cm)	Height of Resulting Box (cm)	Volume of Resulting Box (cm³)
1	14	28	1	392
2	12	26	2	624
3	10	24	3	720
4	8	22	4	704
5	6	20	5	600
6	4	18	6	432
7	2	16	7	224
8	0	14	8	0

Side Length of Cut-Out Square (cm)	Length of Resulting Box (cm)	Width of Resulting Box (cm)	Height of Resulting Box (cm)	Volume of Resulting Box (cm^3)
2	12	26	2	624
2.2	11.6	25.6	2.2	653.312
2.4	11.2	25.2	2.4	677.376
2.6	10.8	24.8	2.6	696.384
2.8	10.4	24.4	2.8	710.528
3	10	24	3	720
3.2	9.6	23.6	3.2	724.992
3.4	9.2	23.2	3.4	725.696
3.6	8.8	22.8	3.6	722.304
3.8	8.4	22.4	3.8	715.008

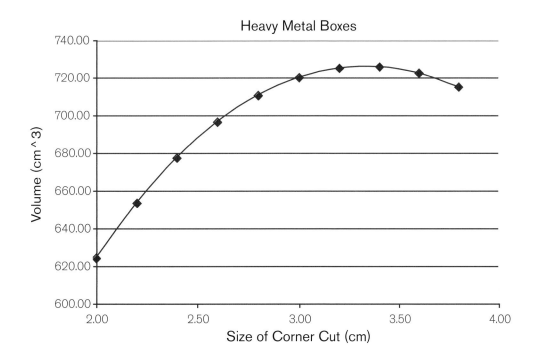

Starting with x = 3 and increasing in increments of 0.05 cm

Side Length of Cut-Out Square (cm)	Length of Resulting Box (cm)	Width of Resulting Box (cm)	Height of Resulting Box (cm)	Volume of Resulting Box (cm³)
3	10	24	3	720
3.05	9.9	23.9	3.05	721.6605
3.1	9.8	23.8	3.1	723.044
3.15	9.7	23.7	3.15	724.1535
3.2	9.6	23.6	3.2	724.992
3.25	9.5	23.5	3.25	725.5625
3.3	9.4	23.4	3.3	725.868
3.35	9.3	23.3	3.35	725.9115
3.4	9.2	23.2	3.4	725.696
3.45	9.1	23.1	3.45	725.2245
3.5	9	23	3.5	724.5

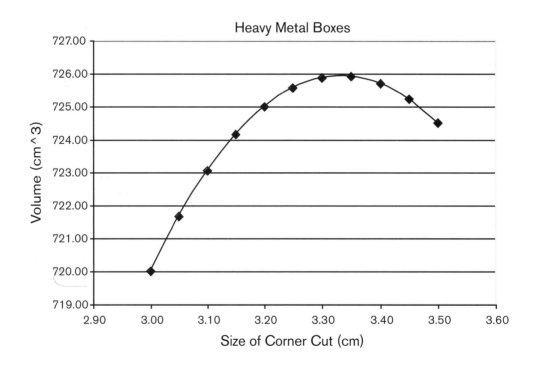

b and *c*. The maximum value that the first table displays for the volume of the box is 720 cm³. The maximum value for volume in the second table is about 725.7 cm³. The maximum value in the third table is about 725.9 cm³. On the basis of these sample tables, the best estimate of the maximum volume of the box is 725.9 cm³.

6. *a.* Students' answers will vary. A comparison of the sample results of narrowing in by graphing the volume function and narrowing in by using successively smaller ranges and increments in a series of tables would lead students to give 725.9 cm³ as their best estimate of the maximum possible volume of the Heavy Metal box.

   b. Students' opinions will vary about which method of narrowing in provides the best estimate of the maximum value. A student might say something like, "I think the tables can give you a better estimate. You can keep finding smaller intervals to test until your largest value no longer increases. This seems to be better than guessing from the graph." By contrast, another student might find that zooming in on a graphing calculator is easier and gives a more precise value for the maximum volume.

7. The third table supports an estimate of 725.91 cm³ as the maximum volume of the box. A box of this volume results when a square of 3.35 cm is cut from each corner of the sheet metal. To support this estimate, students might point to the corresponding graph as evidence (see the graph in the solution to step 5(*a*) above). The graph of $y = x \cdot (16 - 2x) \cdot (30 - 2x)$ has a "peak" between 3.3 and 3. It shows that the slope is decreasing and is virtually 0 at every point near this approximate maximum. This graphical evidence suggests that the absolute change in y-values between the approximate maximum point (3.35, 725.91) and the true maximum point in the interval is less than the change in y-values between (3.35, 725.91) and (3.3, 725.87) or between (3.35, 725.91) and (3.4, 725.70). Thus, a reasonable estimate of the uncertainty, or maximum error, is the larger of these, which is 725.91 − 725.70, or .21. Thus, the maximum should be in the interval 725.9 ± 0.2 cm³.

Solutions for "Not Only Useful, but Also Orna-Metal"

Maximizing the Measurement − Part 2

1. Students' predictions about the dimensions of the box top of largest area will vary. Some students might guess correctly that a square with a side of 25 centimeters will make the box top of maximum area, since making one side very long starts to make the area go to zero.

2. *a−b.* The students should recognize that the first fold must occur between 0 and 50 centimeters since the second fold must occur exactly at the 50-centimeter point. Once someone chooses the position of the first fold, the length of the trim to the second fold is 50 cm minus the length of trim to the first fold. Thus, determining the length to the first fold determines the length and width of the rectangle and hence its area. If the length of the trim to the first fold is x, then the length to the second fold in 50 − x, and the area of the enclosed rectangle is $x(50 - x)$.

3. *a−b.* After exploring various ranges, the students should arrive at a precision of 0.1 cm and a range like that in the following table.

Width of Box Top (cm) (Length of Trim to First Fold)	Length of Box Top (cm) (Length of Trim between First and Second Folds)	Area of Rectangle (cm²)
24.5	25.5	624.75
24.6	25.4	624.84
24.7	25.3	624.91
24.8	25.2	624.96
24.9	25.1	624.99
25.0	25	625.00
25.1	24.9	624.99
25.2	24.8	624.96
25.3	24.7	624.91
25.4	24.6	624.84
25.5	24.5	624.75

The students should observe that the maximum area appears to be 625 cm². As x changes, the change in the value for the area gets smaller near the maximum. Thus, the difference between 625 cm² and the true value should be less than the difference between 625 cm² and the next closest values in the table, or .01. Thus, the students can estimate that the true maximum will be in the interval 625 ± .01.

4. *a.* The equation for the rectangular area of the box top as a function of x is $A(x) = x(50 - x)$.

   b–c. Zooming in on the graph will yield an area of 625 cm² as a rounded-off estimate of the maximum. Again, the uncertainty can be estimated by the difference between the maximum zoom value and the values immediately to the left and right of it on the graph, as given by the "**trace**" option of the calculator. On the zoom that we performed, tracing gave the following sequence of values near the maximum trace value: 624.972, 624.997, 624.997, 624.975. The difference between 624.997 and 624.972 is .025. Rounding to two decimal places, the uncertainty is roughly .03, since the slope of the function is getting closer to 0 in the intervals represented by these trace values. Rounding the maximum value 624.997 to the same decimal place, we can say that the true maximum should be in the interval 625 ± .03.

5. *a.* The students should observe that both methods are effective, and both should appear to be converging to a maximum of 625 cm².

   b. Students should note the symmetry in the table of values and in the graph. The area function is a parabola and the maximum value occurs at its vertex. Students can argue mathematically that the vertex is the point (25, 625). One argument is that the area is 0 at $x = 0$ cm and at $x = 50$ cm. By symmetry, the x-coordinate of the vertex must be exactly halfway between these two x-values. Students may also know that the formula for the x-coordinate of the vertex of a parabola $ax^2 + bx + c$ is $x = \dfrac{-b}{2a}$. This also yields $x = 25$ in the problem under consideration.

c. The students can make the following generalization: The rectangle of maximum area that can be enclosed by a given perimeter is a square whose side length is one-fourth of the perimeter.

6. Students' responses will vary. Students can use tables, graphs, and functions, or they can simply argue from the generalization that they made in step 5(*c*) that the smallest perimeter to area ratio for a rectangle occurs when the region is a square. Thus, the minimum perimeter is $4 \times \sqrt{900}$, or 120, cm. The discussion of the activity in the text (see pp. 81–82) provides other possible responses.

References

D'Angelo, John P., and Douglas B. West. *Mathematical Thinking: Problem-Solving and Proofs*. 2nd ed. Upper Saddle River, N.J.: Prentice Hall, 2000.

Chapman, L. J., and J. P. Chapman. "Atmosphere Effect Re-examined." *Journal of Experimental Psychology* 58 (1959): 220–26.

Foster, Alice, Robert L. Devaney, and Jonathan Choate. *Fractals: A Toolkit of Dynamics Activities*. Emeryville, Calif.: Key Curriculum Press, 1999.

Hanna, G. "Proof, Explanation and Exploration: An Overview." *Educational Studies in Mathematics* 44 (2000): 5–23.

Kelly, Gerard W. *Short-Cut Math*. New York, N.Y.: Dover Publications, 1984.

Martin, Tami. "Fracturing Our Ideas about Dimension." *NCTM Student Math Notes* (November 1991): 1–4.

Martin, Tami S., Sharon M. Soucy McCrone, and Cynthia A. Pulley. "What Our Students Really Know about Proof and Reasoning in Geometry: A Look at Classroom-Based Research Data." Paper presented at the Annual Meeting of the National Council of Teachers of Mathematics, Chicago, Ill., 2000.

National Council of Teachers of Mathematics (NCTM). *Principles and Standards for School Mathematics*. Reston, Va.: NCTM, 2000.

Peitgen, Heinz-Otto F., Hartmut Jürgens, and Dietmar Saupe. *Fractals for the Classroom*. New York: Springer-Verlag, 1992.

Polya, George. *How to Solve It: A New Aspect of Mathematical Method*. Princeton, N.J.: Princeton University Press, 1945.

Shaughnessy, J. Michael, Gloria Barrett, Rick Billstein, Henry A. Kranendonk, and Roxy Peck. *Navigating through Probability in Grades 9–12*. Principles and Standards for School Mathematics Navigations Series. Reston, Va.: National Council of Teachers of Mathematics, 2004.

Supplemental Reading

Bonsangue, Martin V., and Gerald E. Gannon. "From Exploration to Generalization: An Introduction to Necessary and Sufficient Conditions." *Mathematics Teacher* 96 (May 2003): 366–71.

Bremigan, Elizabeth George. "Is It Always True? From Detecting Patterns to Forming Conjectures to Constructing Proofs." *Mathematics Teacher* 97 (February 2004): 96–100.

Conway, John H., and Richard K. Guy. *The Book of Numbers*. New York, N.Y.: Springer-Verlag, 1996.

Cox, Rhonda L. "Using Conjectures to Teach Students the Role of Proof." *Mathematics Teacher* 97 (January 2004): 48–52.

Dahlke, Richard, and Robert Fakler. "Geometric Probability." In *Teaching Statistics and Probability*, 1981 Yearbook of the National Council of Teachers of Mathematics (NCTM), edited by Albert P. Schulte, pp. 143–53. Reston, Va.: NCTM, 1981.

Erickson, Timothy E. "Connecting Data and Geometry." *Mathematics Teacher* 94 (November 2001): 710–14.

Galbraith, Peter. "Mathematics as Reasoning." *Mathematics Teacher* 88 (May 1995): 412–17.

 Galindo, Enrique. "Assessing Justification and Proof in Geometry Classes Taught Using Dynamic Software." *Mathematics Teacher* 91 (January 1998): 76–82.

Garfield, Joan B., and Gal, Iddo. "Teaching and Assessing Statistical Reasoning." In *Developing Mathematical Reasoning in Grades K–12*, 1999 Yearbook of the National Council of Teachers of Mathematics (NCTM), edited by Lee V. Stiff, pp. 207–19. Reston, Va.: NCTM, 1999.

 Herbst, Patricio G. "Engaging Students in Proving: A Double Bind on the Teacher." *Journal for Research in Mathematics Education* 33 (May 2002): 176–203.

 Hodgson, Ted, and Kate J. Riley. "Real-World Problems as Contexts for Proof." *Mathematics Teacher* 94 (December 2001): 724–28.

 Knuth, Eric J. "Proof as a Tool for Learning Mathematics." *Mathematics Teacher* 95 (October 2002): 486–90.

 Knuth, Eric J., and Rebekah L. Elliot. "Characterizing Students' Understandings of Mathematical Proof." *Mathematics Teacher* 91 (November 1998): 714–17.

Martin, George E. *Transformational Geometry: An Introduction to Symmetry.* New York, N.Y.: Springer-Verlag, 1982.

Miller, William A. "Recursion and the Central Polygonal Numbers." *Mathematics Teacher* 48 (December, 1991): 738–46.

 Perham, Arnold E., C.S.V., Bernadette H. Perham, and Faustine L. Perham. "Creating a Learning Environment for Geometric Reasoning." *Mathematics Teacher* 90 (October 1997): 521–26; 542.

 Scher, Daniel. "Dynamic Visualization and Proof: A New Approach to a Classic Problem." *Mathematics Teacher* 96 (September 2003): 394–98.

Teppo, Anne R., and Ted Hodgson. "Dinosaurs, Dinosaur Eggs, and Probability." *Mathematics Teacher* 94 (February 2001): 86–92.

Watson, Jane M. "Statistics in Context." *Mathematics Teacher* 93 (January 2000): 54–58.

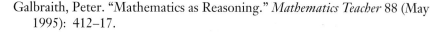